Neurobiological Bases of Abnormal Aggression and Violent Behaviour

József Haller

Neurobiological Bases of Abnormal Aggression and Violent Behaviour

Springer

József Haller
Department of Behavioral Neurobiology
Institute of Experimental Medicine
Budapest, Hungary

ISBN 978-3-7091-1267-0 ISBN 978-3-7091-1268-7 (eBook)
DOI 10.1007/978-3-7091-1268-7
Springer Wien Heidelberg New York Dordrecht London

Library of Congress Control Number: 2014934422

Printed on acid-free paper

Springer is part of Springer Science+Business Media (www.springer.com)

Preface

Understanding the brain control of aggression was greatly enhanced over the last decade. Out of the multitude of new developments, this book will primarily focus on two: the emergence of abnormal aggression models in animals and the introduction and expansion of brain imaging techniques in humans. Although apparently unrelated, these developments bridged an important gap between animal and human aggression research.

In sharp contrast with human research where aggression-related psychopathologies are of major interest, animal research was for long dominated by laboratory models of natural aggressiveness. Newly developed models detached from this classical approach by mimicking etiological factors of aggression-related psychopathologies and by identifying deviant forms of aggression that arise as a consequence of these treatments. Thus, animal research gradually shifted from studying aggression as a natural phenomenon to studying models of abnormal aggression. Recently developed models made clear that laboratory equivalents of etiological factors induce profound changes in brain function and alter qualitatively the neural circuits that are activated by social conflict and that regulate aggressive behavior under these circumstances.

Before the advent of brain imaging techniques, human research lacked adequate tools to study brain function. This is in sharp contrast with animal research where a wide range of advanced technologies were developed over the last decades. In lack of appropriate research tools, assumptions on human brain mechanisms were based on animal studies that were occasionally cross-checked by studies involving people with accidental brain lesions and by brain stimulation techniques that were rather infrequently employed. Brain imaging techniques provide powerful tools of investigation, which showed spectacular advancements over the last decade. New technologies not only allow the study of structural features but also reveal brain connectivity, neurotransmitter receptor profiles, and other important aspects of brain function.

In brief, models of abnormal aggression made animal studies more relevant for human research, while brain imaging techniques made human research more relevant for understanding brain mechanisms. These newly acquired features of the two main sides of aggression research have the potential to make them closer partners and to start a process of integration that is beneficial for both sides and for aggression research in general. The main aim of this book is to create a platform for interactive thinking. We strongly believe that the control of aggression in general and of abnormal aggressiveness in particular cannot be understood without integrating animal findings of translational value and human findings. Close interactions between the two sides offer human research the chance to take advantage of the depth of analysis allowed by the advanced neuroanatomical and neurofunctional methodologies currently used in animal research, while research on animal aggression will greatly benefit from incorporating information on the phenomenon it models. The two developments discussed here opened new channels of information exchange and created the possibility of jointly addressing the triple linkage of etiological factors, brain changes, and behavior and to unravel novel treatment opportunities that take into account all three. This book intends to be the first step towards the integration of concepts and findings that makes this possible.

Budapest, Hungary József Haller

Contents

Chapter 1
Normal and Abnormal Aggressions: Definitions and Operational Approaches

The Latin original of the term aggression ("aggressio") had a military connotation in ancient times and meant "undertaking a task" and "assault." Its psychological and psychiatric meaning was acquired during the twentieth century; the definition "hostile or destructive behavior" was first recorded in 1912 in A. A. Brill's translation of Freud.[1] Hostile and destructive behavior, however, can be understood and measured in different ways especially when the term is used to denote behaviors expressed by different species.

Understanding the neural control of a behavior necessitates the clear delimitation of the behavior that is controlled. While definitions on aggression are in no short supply, these often lack aspects that are crucial for this book: the cross-species validity of definitions, their corroboration with methods of measurement, and the clear delimitation of normal and abnormal forms. To address these issues, here we circumscribe aggressive behavior in two meanings, namely, in theoretical and practical terms. Firstly, we will review intensional definitions that specify the necessary and sufficient conditions to differentiate this behavior from others. This will be followed by extensional definitions that list its main forms and subcategories. While these theoretical constructs clearly demarcate aggressive behavior, they cannot be directly used in research because they do not provide means to measure it. These means are provided by more practical—operational—approaches, particularly by the methodologies that are employed in aggression research to quantify aggressiveness. Animal and human methodologies will be evaluated comparatively; the boundaries between normal and abnormal aggressions will be in focus throughout. The last section of this chapter examines the area of validity of conclusions that are based on particular operational approaches and their relevance for understanding aggression control.

[1] http://www.etymonline.com/index.php?term=aggression

J. Haller, *Neurobiological Bases of Abnormal Aggression and Violent Behaviour*,
DOI 10.1007/978-3-7091-1268-7_1,

1.1 Demarcating and Characterizing Aggression: Theoretical Definitions

1.1.1 Basic Definitions

In principle, there are two types of definitions for "aggression": one focuses on external appearance (i.e., the "form" of behavior), while the other focuses on its functions. "Behavioral form" or "phenomenological" definitions—highly preferred in the 1960s and 1970s—are usually restrictive but have the advantage of being simple and easily applicable to both animals and humans. In plain terms, aggressiveness can be defined as a behavior that delivers noxious stimuli to another organism (Buss 1961) or somewhat more explicitly: any behavior directed towards the goal of harm and injury (Baron 1977; Zillman 1979). These clear-cut definitions immediately raise a simple but relevant question: why should one organism deliver "noxious stimuli" to another? The explanation is provided by the other type of basic definitions, which attribute functions to this behavior. Functional definitions are likely based on the common observation that aggressiveness is present in any species that is endowed with the capacity of performing it, from invertebrates to humans.[2] One cannot reasonably assume that aggressive behavior survived evolutionary pressures for hundreds of millions of years in so many species without providing some functional advantages. It is generally believed that aggression is the most efficient form of competition that bridges the gap between the limitedness of resources on one side and the individual needs of survival and the more general "evolutionary need" of leaving as many descendants as possible on the other side. Thus, the proximate function of aggression is resource competition that ensures survival, while the ultimate function is reproduction that ensures contribution to the gene pool of the next generation. This dual goal is achieved by fights over food, territory, mating partners, etc., and by protecting offspring by aggressive means. This utilitarian view on aggression was formulated very early (Darwin 1871) and was maintained ever since (see Archer 2009; Lindenfors and Tullberg 2011; Weiger and Bear 1988 as examples).

Thus, individuals harm each other to get access to resources and to ensure gene transmission over generations in a world where resources are limited and the delivery of harm is more efficient than negotiation.[3] At the first sight, the functional definition applies to animals only, because survival and the ability to leave progeny rarely depend on aggression in humans. Human societies consist of highly interdependent individuals, which are expected to negotiate rather than fight for resources. Still, human aggression is often gain oriented; this type of human

[2] For arthropods see Kravitz and Huber (2003); for lower animals see van Staaden et al. (2011) and Evans (1973).

[3] Negotiation and sharing are viable options when resources are sufficient for all. This is not always the case.

behavior is commonly called instrumental aggression (Feshbach 1971). Moreover, particular forms of aggression (usually considered legitimate) serve survival in the form of self-defense, the defense of offspring, or the defense of others (Archer 2009; Björkqvist et al. 2000). Furthermore, intricate forms, e.g., relational and indirect aggressions, are means of social competition that are employed at a wide scale in human societies. Because human aggression may become a matter of survival and inclusive fitness under certain conditions, one should attribute functionality to human aggression as well. Taken together, these considerations show that (1) aggression is omnipresent in arthropods and vertebrates and is shown by all lower animals that are endowed with the capacity of performing it and (2) it confers competitive advantage in animals and has similar roles in people under particular conditions.

A final clarification that intensional definitions need refers to the word "harm." In his influential work on aggression, Lorez (1963) notes that "crude" forms of aggression (i.e., the delivery of injuries) gradually progress into "symbolic" actions that fulfill the function of communication. The reason is that aggressiveness involves considerable risks for both the recipient and perpetrator. Either of the fighting parties can suffer injuries, and both lose energy reserves that may become vital in a subsequent challenge (e.g., predator attack) (Haller 1995). Therefore, injurious behaviors not only are gradually replaced with but are also *preceded* by less risky behaviors, e.g., threat signals that offer the chance of withdrawal before the actual fight is started. Such forms of behavior are integrant parts of the aggressive repertoire of animals (see van Staaden et al. 2011 for a review). Human aggression also entails threats, e.g., threatening facial expressions and body language, verbal warnings, etc. It is generally considered that threatening with physical harm is an aggressive act per se. On the other hand, harm should not always be understood in a physical sense. Manipulating social environments (e.g., by malicious gossiping, ignoring, and coercing) may hurt a person in a nondirect manner and may ultimately disrupt his/her social status (Bjorkqvist and Niemela 1992; Coccaro et al. 1997a, b; Galen and Underwood 1997; Gomes 2007; Paquette and Underwood 1999). Social manipulation is not uncommon in animals; the main behavioral tool is social exclusion in their case (see Sasaki and Uchida (2012) for a review and Goodall (1986) for studies in chimpanzees). Therefore, a more complete definition should make reference not only to harm and injury but also to threatening with these; in addition, it should consider "harm" in both physical and psychological terms (Björkqvist et al. 2000; Brower and Price 2001). Further specifications of the word "harm" include intentionality (to exclude accidental events) and exclude benefits for the victim (Berkowitz 1993, 1994; Björkqvist et al. 2000; Bushman and Anderson 2001; Feshbach 1971). As such, aggression includes neither accidental harm nor harm inflicted for positive ends (e.g., to discourage social parasites, to discipline offspring, or to maintain cooperative behavior; Clutton-Brock and Parker 1995). To avoid an impossibly complex sentence, we define aggression below by pointing out its essence and adding modifiers.

Aggression is a behavioral means of resource competition that is based on the delivery of or threatening with harm. Harm should be understood in both physical and psychological terms, should be intentional, and should benefit the victim neither directly nor indirectly.

1.1.2 Definitions for Subtypes

The definition adopted above shows that aggression is not a unitary phenomenon and cannot be the object of studies in such general terms. Therefore, aggression was divided into several subtypes along a variety of classification principles. The form of aggression is one such principle; in humans, subtypes are called physical, verbal, and indirect aggressions (Bjorkqvist and Niemela 1992; Gomes 2007; Paquette and Underwood 1999; Vaillancourt and Sunderani 2011). Physical aggression involves physical harm; verbal aggression (including body language) covers threats that aim at intimidating partners, while indirect aggression consists of social manipulation that inflicts psychological harm in most cases. In some papers, indirect aggression was restricted to or included aggression towards objects (Coccaro et al. 1997a, b; Crowe 1974). This approach was considered outdated by papers published later (Gomes 2007). We suggest here that aggression against objects (e.g., slamming doors) should be considered part of "verbal" aggression when its aim is intimidation and displaced aggression when it serves to channel off emotions. This issue will be detailed below. Rodent analogs of these behavioral categories include biting attacks (physical aggression), offensive threats (verbal aggression), and social exclusion (indirect aggression). Phenomenologically delimited subtypes can be graded according to their precise forms and the severity of consequences. For example, fighting with bare hands is considered milder than weapon use; injuring is a lesser form of aggression than killing (Björkqvist et al. 2000; Holm 1982). Similarly, body language and verbal threats may be perceived as graded versions of "verbal" aggression. Finally, indirect aggression can be divided into direct (conspicuous) and hidden forms (Gomes 2007).

The reason and emotional background of aggression can also serve as criteria for typifying this behavior. Emotional and instrumental forms are the most widely used categories (Blair 2001; Feshbach 1971; Kempes et al. 2005; Lopez-Duran et al. 2009; Raine et al. 2006; van Bokhoven et al. 2005; van Honk et al. 2010). The former (often called reactive) consists of uncontrolled aggressive outbursts that are elicited by perceived threat or provocation. This type of aggression is characterized by high levels of bodily arousal. Instrumental aggression (also called proactive) is controlled and goal directed. It is usually associated with low levels of arousal. Emotional and instrumental aggressions are associated with different types of criminal aggression and aggression-related psychopathologies, as discussed below. There are several rodent models of abnormal aggression, where important aspects of emotional (reactive) and instrumental (proactive) aggression were mimicked (see below). A specific form of emotional responses is displaced aggression; this behavior is directed towards a third party under conditions of frustration and distress. While this behavior is commonly studied in people (Denson et al. 2006; Fitz 1976; Kaufmann and Feshbach 1963; Pedersen et al. 2000; Reijntjes et al. 2007), similar studies in animals are sparse although displaced aggression is readily shown by vertebrates from fish to rats (Overli et al. 2004; Nephew et al. 2003; Hoffmann et al. 1987).

Other frequently used categories are overt and covert aggressions. These may be perceived and defined in different ways. Popma et al. (2007a), for instance, use the phrase "overt aggression" as a synonym of physical and verbal aggressions combined, while "covert aggression" depicts emotional components in their terminology (e.g., hostility, angriness, jealousy, etc.). In other papers, "overt aggression" includes physical fights, weapon use, cruelty, robbery, and sexual abuse, while "covert aggression" is defined as housebreaking, steeling, fire setting, lying, truancy from home, etc. (van Bokhoven et al. 2005). The phrases "overt" and "covert" are sometimes perceived as synonyms of physical and indirect aggressions, respectively (Vaillancourt and Sunderani 2011).

The brief review of various classifications shows that the terms are not used consistently, and there are overlaps between various categories. For example, indirect aggression is instrumental in many cases; covert aggression is considered synonymous with hostility and anger according to one categorization and with nonviolent lawbreaking according to the other. Any combination is possible between emotional (reactive) and instrumental (proactive) aggressions on one side and physical and verbal aggressions on the other. In addition, such subcategories do not categorize perpetrators but actions: one and the same person may show emotional aggression under certain circumstances and instrumental aggression under others. Despite inconsistencies and overlaps, these concepts are very useful for differentiating and studying distinct classes of behaviors.

Another criterion that may be used to typify aggression relates to its normality. Given the importance of this criterion for the present work, it will be presented in a separate section.

1.1.3 Normal and Abnormal Aggressions: Basic Principles

As shown above, aggression is ubiquitous in animals where it plays important functions. For this reason, abnormality may be perceived unlikely as far as animals are concerned. On the other hand, the high level of socialization and the strong moral, legal, and social stigmatization of aggressiveness may render aggression per se abnormal in humans. In simple terms, one may be tempted to believe that aggression is always normal in animals and is always abnormal in people. Neither assumption seems true.

Conflicts in animals are settled according to certain rules, which result from evolutionary pressures against dangerous forms of competitiveness (Hinde 1974). Intraspecific aggression is not aimed at inflicting lethal injuries but at ensuring access to resources. Species would be rapidly eliminated if its individuals risked their lives to obtain food, mating partners, etc. Moreover, the near-emptying of biochemical energy stores by excessive fighting per se may put the life of animals at risk (Haller 1995). Evolution shaped aggression such that these risks are avoided. Aggressive intent is signaled in advance to allow withdrawal, the level of aggressiveness is kept between certain limits to minimize risks and excessive energy losses, attacks are targeted towards body parts that lack vital organs and are not excessively exposed to infections, aggressiveness is rapidly deferred by submission or flight by the opponent, and, finally, females (rate-limiting factors in reproduction) are largely protected from attacks by males (not considering sexual coercive acts). Deviations from these rules capsize the balance between costs and benefits, render aggression nonfunctional, and provide a handle for differentiating normal and abnormal aggressions in animals (Haller et al. 2001, 2005a; Haller and Kruk 2006). Thus, abnormal aggression is defined in terms of natural rule breaking and the lack of functionality in animals.

Human aggression is considered abnormal if it violates moral and legal rules or if is associated with psychopathologic conditions. These two principles (morality, law and mental health) are closely related. An interesting study by Björkqvist et al. (2000) demonstrated that people base their attribution of aggression on factors that are consistent with the severity of punishment in sentences for different crimes of violence; thus, moral and legal thinking is closely related as far as aggression is concerned. Significant overlaps were also found between the behavioral and legal correlates of aggression as well as between aggression-related psychopathological constructs and criminal behavior (Coccaro and McNamee 1998; Lee and Coccaro 2001; Gao and Raine 2010; Yang et al. 2008). Moreover, the characterization of aggression-related psychopathologies by DSM-IV-TR explicitly includes rule breaking among the symptoms of conduct, oppositional-defiant, and antisocial personality disorders and implicitly in the case of others (e.g., intermittent explosive disorder) (American Psychiatric Association 2000). Overall, this source considers psychological conditions pathological if they inflict suffering; cause significant impairment in social, occupational, or other important areas of functioning; and are persistent. Persistent violent criminal behavior fulfills these criteria.

In legal terms, the key factor that differentiates normal and abnormal aggressions is rule breaking. For example, violence was recently defined as the infliction of physical harm "*in violation of social norms*" (Brower and Price 2001). In psychopathological terms, aggression is abnormal if it makes the individual dysfunctional. Persistent aggressiveness leads to social exclusion, to the loss of jobs, to destruction of families, etc.—briefly, it imperils social, occupational, and other important areas of functioning. Thus, legal and psychopathological delimitations of human abnormal aggression, when taken together, replicate the definition of abnormal aggression proposed above for animals; human abnormal aggression can be defined in terms of rule breaking and dysfunctionality. Differences between animals and humans lie in details only. For example, rules are shaped by evolutionary pressures in animals, while human rules are formalized. The nature of punishment may also be dissimilar; abnormally aggressive animals may take punishment in the form of defeats when meeting opponents of superior fighting abilities, while humans risk the same and legal punishment in addition. Even the very long-term consequences appear comparable; excessive increases in the share of abnormally aggressive individuals imperil the success of animal species on an evolutionary scale; in the case of humans, the same jeopardizes the success of societies on a historical scale. Given these close parallels, we define below abnormal aggression by summarizing all the considerations outlined above.

Abnormal aggression breaks natural or formalized rules and is dysfunctional on the long run in terms of negative consequences for the perpetrator. This definition implies that certain forms of aggression are not abnormal in humans. Indirect aggression, for instance, does not break the law and may reach its function if carried out skillfully (i.e., it may confer competitive advantage). The same holds true for verbal aggression in many instances. Even physical aggression may be normal if performed by authorized bodies (e.g., law enforcement) or if lawful (e.g., self-defense). On the other side, some forms of aggression may be abnormal in animals. In the laboratory, such abnormal forms can be induced by treatments that mimic etiological factors of human abnormal aggression. Although the issue was not investigated systematically, the conditions leading to abnormal aggression (e.g., excessive stressors suffered in certain periods of life; see below) are likely occurring under natural conditions, which makes it probable that abnormal aggression exists in free-ranging populations.

1.1.4 A Theoretical Classification of Aggressive Behaviors

We suggest here a classification of aggressive acts based on the constructs overviewed above. The classification was visualized in Fig. 1.1. This figure will be used as a template for evaluating the correspondences between theoretical definitions and operational approaches. Animal aggression was not considered at this stage, but will be included in the system later on. We believe that there are three principles along which aggression should be categorized: behavioral form, reason, and abnormality.

The forms of behavior were listed in the rows of the figure. The three main categories mentioned above (physical, verbal, and indirect) were slightly changed. Particularly, the phrase "verbal aggression" was replaced with "intimidation" to accommodate nonverbal forms that have the same function: body language and object-directed aggression when the latter is displayed with the aim of intimidation (e.g., hammering the table, slamming doors, etc.). Other forms of object-directed aggression (e.g., breaking objects to channel emotions) are considered displaced aggression. Instead of the word "indirect aggression," we use here "relational aggression" to avoid the inconsistencies mentioned above; the two phrases are often used as synonyms in the literature. The severity of the behaviors increases downwards; intimidation is considered the least severe form because it only threatens with harm; relational aggression involves harm in psychological meaning,

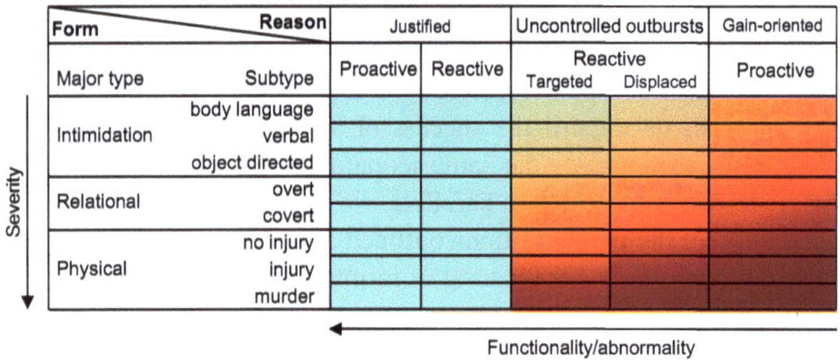

Fig. 1.1 The classification of aggressive acts based on behavioral characteristics (*row headings*), type/reason (*column headings*), and the level of functionality/abnormality (represented by *color code*; functionality decreases from *blue* to *dark red*). For details see Sect. 1.1.4

while physical aggression in physical meaning. These major forms were also graded according to more specific behavioral aspects. Body language was considered a milder form of intimidation than verbal aggression or object-directed aggression. In the case of indirect aggression, overt forms were considered less severe because they allow for mitigation. The level of injury infliction differentiates physical forms of aggression.

The reasons of being aggressive were listed in column headings. The following reasons were differentiated: legitimate uses of aggression (e.g., the defense of others and self-defense), the inability to control emotions and behavior (uncontrolled outbursts of anger), and gain-motivated aggression. Legitimate uses of aggression were further subdivided into proactive and reactive forms. The former is typical to law-enforcement bodies, while the latter is usually shown by those defending themselves from attack. Uncontrolled outbursts of anger were further subdivided into target oriented (e.g., aimed at the source of threat or provocation) and displaced (channeled towards objects or a third party). The legitimacy of the three main reasons decreases from left to right. Any form of defense is considered legitimate provided that it is proportional; disproportional aggression belongs to one of the subsequent two categories. Uncontrolled outbursts of anger cover aggressive acts that are not justified by the situation (i.e., they are out of proportions). This form of aggression is usually called reactive/emotional and is characterized by strong emotional reactions. Such actions are not legitimate, even less so if the target is a third party (i.e., displaced aggression against people). Yet in legal and moral terms, acting under psychological distress is extenuative to a certain extent; therefore, reactive/emotional aggression was considered less rule breaking (somewhat more legitimate) than instrumental aggression. The least legitimate forms of aggression are those performed for gain.

The color code of Fig. 1.1 denotes abnormality based on the functionality of behavioral acts. As aggressiveness is believed to be a means of competition, functionality was evaluated from the point of view of competitive advantages. On the short term, any of the aggressive behaviors listed in Fig. 1.1 may be considered functional according to this criterion. However, certain types of aggressive behaviors are competitively disadvantageous on the long run, for which these were classified as dysfunctional. For example, excessive and uncontrolled outbursts of aggression disrupt families, social networks, and work places, which renders this behavior dysfunctional overall, even if such outbursts may serve momentary interests in terms of assertiveness under particular circumstances. Similarly, instrumental aggression may make the individual successful in certain competitive situations, but on the long run, they involve personal risks and social punitive reactions that make instrumental aggression dysfunctional overall. By contrast, defense-oriented aggression is advantageous on the short run (it averts dangers) and does not entail long-term risks in legal or other terms. Therefore, defensive aggression was considered functional. Social acceptability and the severity of aggressive acts were also taken into account to grade the dysfunctional nature of aggressive behavior. Intimidation and relational aggression are ubiquitously used by humans as means of competition. Expressing displeasure by body language, verbally, or by object-directed aggression (e.g., slamming doors, hammering tables) is quite common during conflict. Spreading rumors, excluding people from social networks, or setting confederates against competitors is similarly common within social groups. Regular involvement in such behaviors may disfavor the individual on the long run, but negative consequences are not as serious as with physical forms of aggression. Therefore, these behaviors were considered less dysfunctional.

1.2 Quantifying Aggression: Operational Approaches

The definitions discussed above are intended to demarcate aggressive from other types of behavior and to separate its subtypes. Neither definition is sufficient per se to study aggression; these require specific methodologies. This chapter describes the main approaches in animals and humans, with a special focus on quantitative measures and the relevance and usefulness of the findings obtained for understanding normal and abnormal aggressions. We mention that each model and approach will be discussed in more detail in the subsequent chapters. Here we provide a general overview only.

1.2.1 Models of Normal Aggression in Animals

In animals, aggression is usually studied "in action," i.e., while aggressive acts are actually performed. This is in sharp contrast with the most frequently employed approaches in human aggression research (see below). Even in animals, however, the neurobiology of aggression is rarely studied under entirely naturalistic conditions, partly because the required equipment and the data sampling procedures need laboratory environments, partly because laboratory studies ensure a stability of conditions that are rarely met in nature. Therefore, aggression is usually studied by employing models. The most fruitful models used so far are brain-stimulation-evoked aggression models and the resident-intruder test (the latter including studies on maternal aggression). Other methodologies include defensive, fear-induced, shock-induced, and social aggressions. Models of predatory aggression also exist; these will be discussed later on in relation with intraspecific aggression only. The methodologies discussed here are assumed to address various aspects of normal aggressiveness, i.e., are supposed to model aggressiveness shown by animals under naturalistic conditions. We will briefly characterize the main models below.

1.2.1.1 The Resident-Intruder Test in Males and Females

In its "classical" version, male subjects (usually rodents) are moved to an individual cage and are faced with an opponent (intruder) after a suitable time, usually after a couple of days. The behavior of rodents in this model was first described by Grant and Mackintosh (1963) and can be considered the most popular aggression model ever since. It is believed that individual housing activates territoriality; as such, aggressiveness elicited by this test is perfectly in line with functional definitions of aggression, because territories are the resources animals are most frequently fighting for under natural circumstances. Aggressiveness is increased by cohabitation with a female; this condition may add to the functional relevance of the model, because reproduction-related aggressiveness is also among the strongest naturalistic motivations. In phenomenological terms, aggressiveness shown in this model consists of various social, offensive, defensive, and dominance-related postures as well as biting attacks. The latter are overwhelmingly targeted towards the back and flanks of opponents (i.e., regions that contain no vital organs and are relatively less prone to infections); in addition, the frequency of biting attacks is reasonable (5–10 bites over a usual test). Thus, the behavior of subjects follows the "rules" outlined above. The latency of attack, the number of biting attacks, and the duration of aggressive postures (threats) are used as measures of aggressiveness. Social behaviors may be studied to complete the picture, while the durations of defensive and dominance-related behaviors may be used to evaluate the outcome of the fight. The contribution of this model to aggression research is substantial and diverse; it was extensively used to elucidate the roles of environmental conditions and hormones; it was used in pharmacology-oriented studies and those focusing on genetic factors, as well as in studies on mechanisms of neurobiological control. The models of abnormal aggression are also based on the resident-intruder test.

Albeit regular resident-intruder tests can be performed in females, aggressiveness is rather low in this gender for which this approach is rarely employed. By contrast, females violently defend pups from unfamiliar males that are prone to kill them. The behavior differentiates targets; female intruders (less dangerous for pups) are rarely attacked. This behavior called maternal aggression is intensely studied and can be considered the "female variant" of the resident-intruder test. The first recorded use of the model dates back to the early 1970s (Gandelman 1972). While aggressiveness shown in the resident-intruder and maternal aggression models is similar in many respects, there are important differences. In contrast to males, females are less tending to signal their attacks by offensive threats and readily attack vulnerable body parts of intruding males (Parmigiani et al. 1988). Albeit this behavioral profile is akin to the one marked as abnormal in males, it seems perfectly functional in this case, because subjects fight opponents that are superior in body size and defend resources of utmost value. An interesting feature of maternal aggression is its association with reduced stress responses that is also seen in certain models of abnormal aggression (see below). Particularly, maternal aggression is associated with low glucocorticoid stress responses (Neumann 2001; Neumann et al. 2001); moreover, acute stressors suffered before the encounter and increased stress responsiveness (as an individual feature) decrease maternal aggression (Gammie and Stevenson 2006; Gammie et al. 2005). A recent study in humans also showed that reduced autonomic arousal disinhibits female aggression in humans (Hahn-Holbrook et al. 2011). Taken together, these findings suggest that maternal aggression—although normal—shows similarities with certain models of abnormal aggression in both behavioral and emotional terms; as such, the maternal aggression paradigm offers unique possibilities to understand the control of aggression.

1.2.1.2 Brain-Stimulation-Evoked Aggression

Although this is not a naturalistic model of aggression, it is believed to be based on the activation of brain mechanisms that underlie natural forms of aggressiveness. As such, this induced behavior may reveal the neurobiological mechanisms of aggression in general; this justifies its presentation in this section. The methodology was introduced by Hess (1928), who observed that the electrical stimulation of particular hypothalamic sites rapidly induces biting attacks directed towards social partners. In the lack of the latter, no behavioral responses are evident. This methodology contributed to aggression research by delimitating aggression-related hypothalamic brain regions and by elucidating the major components of aggression-related neural networks (for reviews see Kruk 1991; Siegel et al. 1999). The measure of aggression is the strength of the electrical stimulation that is required to elicit aggression, which is inversely related to aggressiveness (larger current thresholds denote suppressed aggression). Networks subserving attacks were usually identified by the parallel stimulation of two or more brain areas and by the combination of stimulation and lesion techniques. The technology is also suitable for studies on the pharmacology of aggression; in this case, stimulation is combined with the systemic or local brain application of pharmacological agents. Electrical stimulation is rarely used today to study aggression, perhaps because it requires sophisticated equipment and deep knowledgc in the field of physics. Another reason may be its low neuroanatomical resolution because electric currents stimulate all the neuron and at least some of the passing fibers that surround the electrode. The place of this technology may be overtaken by optogenetics which has a considerably better neuroanatomical specificity (Lin et al. 2011). It is important to note that brain-stimulation-evoked aggression fulfills most criteria based on which normal and abnormal aggressions are differentiated. Vulnerable targets, females, submissive animals, etc., are readily attacked, and the behavior is not functional by any means, because it is motivated by stimulation, not by the social situation. Nevertheless, it reveals the basic mechanisms of aggression, and the neural networks identified with this methodology guided research made in more naturalistic models.

1.2.1.3 Shock-Induced Aggression

It was observed rather early that the administration of mild electric shocks immediately elicits aggressiveness in a novel environment where spontaneous aggression is rare (Eichelman 1971; Powell et al. 1969). Albeit the model gained some popularity in the 1970s and 1980s and is still in use today, it did not have an impact comparable to the two models discussed above. A PubMed search with the search terms (rat [title/abstract] OR rats [title/abstract] OR mouse [title/abstract] OR mice [title/abstract] AND aggression [title/abstract] AND shock-induced [title/abstract]) returned 76 hits, while the same search term without the phrase "shock-induced" returned over 2,300, suggesting that the model is not used very frequently. The likely reason is the strange behaviors elicited by shocks. Unfortunately, the studies employing the model tend to provide global scores for aggressiveness, for which it is difficult to evaluate what actually happened during the test. However, "aggressiveness" elicited by shocks is a mixture of offense and defense according to our personal experience. Biting attacks are rare.

1.2.1.4 Fear-Induced Aggression

Models of this type involve the exposure of subjects to conspecifics, predators, or humans under conditions that elicit fear (*conspecifics*: Blanchard and Blanchard 1989; *predators*: Blanchard et al. 1990; *humans*: Plyusnina and Oskina 1997). The models in use are centered on the notion of defensive aggression, supposed to be different from offensive aggression. Such models are used by a restricted number of research groups only. Reasons may be multiple and may range from a faulty understanding of the issue by the aggression research community to practical and theoretical problems associated with such models. To name a few of these: (1) offense and defense are always intermixed irrespective to the status of the individual; (e.g., both residents and intruders show offense and defense in the resident-intruder test), (2) situational and behavioral defensiveness and offensiveness are also intermixed (e.g., residents defend their territory by offensive behaviors), and (3) strong defensive responses (e.g., those elicited by predators) are so much mixed with fear that they were proposed as models for various types of anxiety (*panic*: Blanchard et al. 1997; *post-traumatic stress disorder*: Hammack et al. 2012). While defensive behaviors are just as important for the individual as offensive ones, and defense may involve acts of offense under particular conditions, they seem to be motivated primarily by fear. As such, these behaviors are only partly consistent with phenomenological definitions and inconsistent with functional ones, for which will receive marginal attention in this book.

1.2.1.5 Social Aggression

The closest analogs of human aggressive behavior are tests where subjects live in colonies, and aggressiveness is studied in conjunction with their social life. The prototypical example of such models is the visible burrow system developed by Blanchard et al. (1988). In this model, rat colonies live together for long periods of time in large enclosures that include burrows (natural living spaces for rodents). Less ambitious models also exist; in their case, the length of cohabitation is restricted to a couple of weeks and the enclosure is smaller and does not include burrows (Mikics et al. 2007; Millard and Gentsch 2006). Colony-based models were used to evaluate the impact of group living on the ontogenetic development of emotional and stress responses (Kaiser et al. 2007) and to study social integration in animals submitted to models of abnormal aggression (Tulogdi et al. 2014). Despite their obvious value for studying aggression in social species (e.g., rats), these models are rarely employed, likely because they are extremely labor intensive and require large laboratory space.

1.2.2 Models of Abnormal Aggression in Animals

The idea of studying abnormal forms of aggression in rodents is relatively new in aggression research: our lab was the first to propose a model (Haller et al. 2001). This does definitely not mean that abnormal aggression was not seen in rodents before this publication; moreover, models used earlier may readily be included into the large family of abnormal aggression models. There were three concepts that differentiated this publication from the earlier ones. First, we deliberately generated the model by making use of data on the etiological factors of antisocial aggressiveness. Second, we detached from the generally employed quantitative approach: instead of counting attacks, we focused on their form and behavioral context. Third, attention was not restricted to aggressive behavior per se; we sought to investigate other features related to antisocial aggressiveness (e.g., autonomic responses, anxiety, and social behaviors beyond aggressiveness) (Haller et al. 2004). Based on the principles underlying this experimental work, we developed a general concept for abnormal aggression models (Haller et al. 2005a; Haller and Kruk 2006). The same complex approach was employed when our second model of abnormal aggression was developed (Toth et al. 2008, 2011).

Models of abnormal aggression gained terrain in two directions: models used earlier were reevaluated in the light of this concept (see Miczek et al. (2013) for a review), and new models were developed to address other etiological factors of abnormal aggression. The criteria that differentiate normal from abnormal aggression were reformulated in several ways by the authors who undertook research in this relatively new field. Nevertheless, the essential points remained the same. Rats are submitted to procedures that mimic etiological factors of aggression-related psychopathologies and are tested for abnormal attack features in the resident-intruder test. These features can be of three types: mismatch between provocation and response, disregard of species-specific rules, and insensitivity towards the social signals of opponents.

The etiological factors considered in abnormal aggression models include the application of stressors at various ages, treatments with drugs, and genetic manipulations. We will review below the main models based on the type of the etiological factor that produced abnormal aggression. Two models will be discussed separately, because of their specific features. One model was based on the repeated exposure of rats to victories, while the other on the suppression of glucocorticoid production. In all cases, the emotional background of aggression will be evaluated, because this feature seems relevant to the type of aggression performed and the classification of the models.

1.2.2.1 Stress Models

Human studies show that stressors suffered in early childhood or adolescence (e.g., abuse, maternal and social neglect, as well as other types of severe stressful life events) considerably increase the risks of violent and antisocial behaviors later in life (Attar et al. 1994; Chapple et al. 2005; Dodge et al. 1990; Loeber and Hay 1997; Paperny and Deisher 1983; Trentacosta and Shaw 2008; Widom 1989; Windle and Windle 1995). In addition, acute provocations elicit immediate increases of aggressiveness in adulthood (Baldwin and Randolph 1982; Pedersen et al. 2000; Archer and Benson 2008).

These findings served to develop a series of abnormal aggression models. Repeated maternal separation was used to mimic early maternal neglect (*rats*: Veenema et al. 2006), postweaning social isolation was used to model early social neglect (*mice*: Bibancos et al. 2007; *rats*: Toth et al. 2008; Wall et al. 2012[4]), early social subjugation was used to model adolescent abuse (*hamsters*: Delville et al. 1998; *rats*: Cunningham and McGinnis 2008), and, more recently, strong physical stressors were repeatedly administered to pubertal rats to mimic the impact of nonsocial stressors suffered at this age (Márquez et al. 2013). In these models, subjects were exposed to the stressors in the life periods indicated above, while

[4] Albeit a similar model was developed earlier in this species, this model detaches in terms of the duration of postweaning social isolation and the age at which aggression was tested (see below).

aggression was studied by the resident-intruder test. Finally, the immediate effects of stressors were mimicked by the "instigation" model of aggression, where mice are exposed to territory intrusion without being able to contact the opponent physically, which is followed by exposure to a second opponent which now is physically accessible (Fish et al. 1999). Noteworthy, models were attributed to authors who were the first to provide data from which abnormal features of aggression could be deduced.

The models listed here are in different stages regarding the characterization of aggressive behaviors that result from stress application. Only disproportionality (excessive attack counts) denotes abnormality in the maternal separation (Veenema et al. 2006) and instigation models (Fish et al. 1999). Increases are dramatic in the latter, which renders this model appropriate for investigating the neurobiological mechanisms underlying the immediate effects of stressors on abnormal forms of aggression. In the maternal separation model, dramatic increases in aggression were noticed in juveniles but not in adults, suggesting that the consequences of this stressor are either washed out during development or that adult behavior needs more detailed analysis (Veenema et al. 2006; Veenema and Neumann 2009). Note that abnormal attack features may be present even in the absence of increased attack counts (see below). Interestingly, maternal separation decreased aggression in mice (Tsuda et al. 2011); unfortunately, however, aggressive behavior was not presented in detail in this paper. Dramatic increases in aggression were noticed in the early subjugation model, where, in addition to this quantitative change, an interesting phenomenon also occurred: while aggressiveness increased when subjects faced smaller opponents, it markedly decreased when opponents faced similar or larger opponents (Delville et al. 1998; Ferris 2003). We suggested that this "offensive ambiguity" may be listed among abnormal attack features (Haller 2013). In rats, early subjugation also increased aggressiveness; this was evident when rats were provoked during aggressive encounters by regularly applied tail pinches (Cunningham and McGinnis 2008). Based on the composite score of aggressiveness provided, the increase was dramatic (four–fivefold). Aggression towards females was not increased in the rat subjugation model (Frahm et al. 2011). Taken together, these data suggest that early subjugation in rats can also be considered a model of abnormal aggression based on the etiological factor and the exacerbated response to provocation.

The study of behavior was more elaborate in other models. Albeit the increase in aggressiveness was not particularly large in either, qualitative changes were dramatic, suggesting natural rule-breaking behavior in the terms discussed above. Pubertal stressors increased attacks on vulnerable targets and, in addition, attacks on anesthetized opponents (Márquez et al. 2013), while postweaning social isolation resulted in behavioral agitation during aggressive conflicts, attacks on vulnerable targets, and decreased signaling of attacks; moreover, these subjects delivered biting attacks from submissive posture (Toth et al. 2008, 2011). Dramatic increases in play-fight-related aggressiveness were seen in rats socially isolated from weaning that were tested in late adolescence (Wall et al. 2012); this suggests that postweaning social isolation affects aggressiveness from early ages till adulthood.

The mouse postweaning social isolation model was less well characterized so far, but mice submitted to this model, in contrast to controls, attacked opponents in neutral arenas, suggesting abnormal behavior.

Stress responses may be relevant for evaluating the emotional component of attacks. In this respect it should be noted that early stressors increased stress responses to aggressive encounters in adulthood in most cases (*maternal separation model*: Veenema and Neumann 2009; *pubertal stress model*: Márquez et al. 2013; *postweaning social isolation model*: Toth et al. 2008, 2011; *instigation model*: Fish et al. 1999). The smallest effect was noticed in the pubertal stress model where only the testosterone/corticosterone ratio increased; the strongest effect was seen in the postweaning social isolation model where both autonomic and glucocorticoid responses to aggression showed dramatic increases. Notably, however, fight-induced increases in plasma glucocorticoids were markedly decreased in the subjugation model (Ferris 2003). Thus, early stressors increase emotional responses to aggression except for subjugation which does the opposite.

1.2.2.2 Drug Models

These models were based on human observations on the effects of alcohol, anabolic-androgenic steroids, amphetamine, and cocaine on aggression (Bennett et al. 1969; Cherek et al. 1986; Licata et al. 1993; Su et al. 1993). The effects of these agents on aggression were studied by a variety of authors (*alcohol, mice*: Miczek et al. 1992; *anabolic-androgenic steroids, hamsters*: Melloni and Ferris 1996; *anabolic-androgenic steroids, mice*: Ambar and Chiavegatto 2009; *amphetamine, voles*: Gobrogge et al. 2009; *cocaine, hamsters*: Ricci et al. 2005). Unfortunately, only attack counts were studied in these models, except for amphetamine, where it was shown that adult voles treated with this compound for 3 days attacked females (Gobrogge et al. 2009). The dramatic increase in attack counts may be considered, however, arguments in favor of a shift towards abnormality. The effects of these treatments on aggression-related stress responses were not studied so far. Based on aggression-unrelated papers, where the conditions of drug administrations were similar, we hypothesized that anabolic steroids, amphetamine, and cocaine may have augmented the stress response to aggression, while alcohol (under the conditions employed by Miczek et al. 1992) may have done the opposite.

1.2.2.3 Genetic Models

Such models include selection lines as well as transgenic mice. Aggression has a strong heritability component (Miles and Carey 1997); therefore, it is not surprising that highly aggressive and very docile mice can be produced within a few generations of selective breading (Benus et al. 1991). The aggressive line (called short attack latency mice) produced by this early breeding program shows abnormal features of aggression in multiple ways. When faced with opponents, they deliver an exceedingly high number of bites to any opponent including females as well as immobilized and submissive male intruders; in addition, a large share of these attacks are targeted towards vulnerable body parts of opponents (Benus et al. 1991; Haller et al. 2006; Natarajan et al. 2009a, b; Sluyter et al. 2003). Other highly aggressive selection lines also exist; albeit these do show heightened attack counts during aggressive conflicts, the patterns of their aggressiveness lack the abnormal features seen in short attack latency mice (Natarajan et al. 2009b). Interestingly, decreased autonomic and glucocorticoid stress responses were seen in short attack latency mice but not in other selection lines, suggesting that such features are typical to individuals showing abnormal forms of aggression (Veenema et al. 2004; Natarajan and Caramaschi 2010).

The inclusion of lines selected for anxiety among the models of abnormal aggression may be surprising at the first sight, but recent studies clearly demonstrate that both high- and low-anxiety rats show abnormal features of aggression. In humans, faulty emotion control is a strong reason to behave aggressively (Davidson et al. 2000); the association of disturbed anxiety-like behavior and abnormal aggressiveness may be analogous to this phenomenon. Compared to the nonselected line, both high- and low-anxiety rats show increased aggressiveness and both attack vulnerable targets; in addition, low-anxiety rats attack females and narcotized opponents (Beiderbeck et al. 2012; Neumann et al. 2010). The high- and low-anxiety lines showed decreased and increased, respectively, glucocorticoid stress responses to fighting (Neumann et al. 2010). This feature was opposite to that expected based on their anxiety profile and was also opposite to that seen in nonsocial situations, suggesting that aggression activated specific neuroendocrine mechanisms in these two lines.

The association between monoamine oxidase A (MAO_A) deficiency and aggressiveness was first shown in a Dutch family where the gene of this enzyme was disrupted (Brunner et al. 1993). A similar association was shown in MAO_A knockout mice shortly thereafter (Cases et al. 1995). Subsequent research amply confirmed that deficient MAO_A functioning is associated with violence and antisocial tendencies in humans. Unfortunately, the aggressive behavior of MAO_A knockouts was not investigated in detail so far; therefore, it is unclear whether this line shows abnormal features other than increased aggressiveness. Disrupted social behavior was found in MAO_A hypomorphic mice where the expression of the enzyme was reduced but not eliminated (Bortolato et al. 2011). These mice, however, were not particularly aggressive. Stress responsiveness was poorly investigated in MAO_A knockouts; a study on the long-term effects of the MAO_A inhibitor moclobemide suggests that glucocorticoid stress responses are decreased by chronic MAO_A deficiency (Reul et al. 1994). No data are available on autonomic responses; in general, MAOA inhibitors lack cardiovascular side effects (Bonnet 2003). Considering their relevance, surprisingly little is known about the aggression-related features of MAO_A knockouts.

Naturally, aggression was altered (sometimes dramatically increased) in a series of other transgenic mice. These will be mentioned later. Nevertheless, we found that there is a weak correspondence between the altered genes in mice and aggression-related gene expression profiles in humans. As such, models other than the MAO_A knockout do not seem to be related to etiological factors of human aggression and will not be discussed here.

1.2.2.4 The Repeated Victory Model

Research in humans revealed that a history of aggression is among the strongest predictors of violence (Amore et al. 2008; Allen and Links 2012; Steinert 2002). The repeated exposure of rats to aggressive conflicts seems to have a comparable effect. The tenth aggressive encounter is marked by decreased attack latencies and increased attack counts as compared to the first one; in addition, rats become aggressive against females and anesthetized intruders and fail to signal their attack intentions by offensive threats. Thus, a history of wins in resident-intruder conflicts results in violent forms of aggression in rats. The phenomenon was described in feral rats that showed high levels of aggression at baseline; more mildly aggressive rats showed no similar changes over time. Unfortunately, the model was presented only in reviews so far (de Boer et al. 2003; Miczek et al. 2013). Albeit these reviews did contain experimental data in the form of graphs, the model cannot be accurately evaluated in the absence of precisely described experiments. One of the reviews suggested that this form of aggression is associated with mildly decreased glucocorticoid responses, while a publication by a different set of authors showed that repeated exposure to fights ameliorates aggression-induced autonomic responses in feral rats (Sgoifo et al. 2001). These findings suggest that the shift towards abnormal forms of aggression is associated decreased emotionality. We believe that this interesting model received much less attention by its developers than it deserves based on its human relevance.

1.2.2.5 The Glucocorticoid Deficit Model

The possibility that the HPA-axis can chronically lower its activity—together with the assumption that this condition leads to psychopathology—was recognized by a study on violent antisocially disordered people (Virkkunen 1985) and received much attention in subsequent studies on abnormal forms of aggression. Although contradictory findings also exist, the association between chronically low glucocorticoid levels and antisocial aggressiveness was repeatedly shown in a variety of human subjects (Fairchild et al. 2008; McBurnett et al. 2000; Ruttle et al. 2011; Sondeijker et al. 2008; van Goozen et al. 1998a). We mimicked this endocrine condition by adrenalectomy combined with low-level glucocorticoid replacement that ensured low but physiological plasma glucocorticoid levels, which naturally were not responsive to environmental stressors. This treatment did not affect overall attack counts, but dramatically increased attacks on vulnerable body parts of opponents (head, throat, and belly) and decreased intention signaling by offensive threats (Haller et al. 2001). Importantly, the effects of chronic decreases in plasma glucocorticoids were qualitatively different from those induced by acute deficiency and could be abolished by repeated but not by single glucocorticoid treatments (Haller et al. 2001, 2004). Chronic glucocorticoid deficiency also decreased fight-related autonomic responses and disrupted social behaviors without affecting anxiety. Thus, mimicking an endocrine condition repeatedly associated with anti-social aggressiveness in humans resulted in rule-breaking behavior in rats; other features associated with antisocial aggression, e.g., low emotionality, were also reproduced (Haller et al. 2005a; Haller and Kruk 2006).

1.2.3 Classification of Abnormal Aggression Models

Based on the features described above, we suggest that there is sufficient information to establish a correspondence between the theoretical definitions of aggression and rodent aggression models. Correspondences were visualized in Fig. 1.2, where we superimposed animal models of aggression over the classification system presented in Fig. 1.1. Proactive aggression seen in maternal aggression was found similar to protective proactive aggression performed by humans. Both behaviors are functional and normal; both are performed on the background of reduced emotions. The resident-intruder test was considered analogous with legitimate reactive aggression. Territorial behavior is common and functional in rats. The same is true for self-defense in humans.

Certain models of abnormal aggression were considered analogous with reactive aggression in humans, while others with proactive aggression. The differentiating principle was emotional background. As shown above, abnormal reactive aggression in humans ("uncontrolled outbursts") is a response to provocation or threat and is associated with excessive emotional and stress responses. By analogy, abnormal aggressions associated with excessive emotional and stress responses were considered similar to this type of human aggression. The best example is the postweaning social isolation model, where subjects are behaviorally agitated during aggressive conflicts and show markedly enhanced glucocorticoid and autonomic stress responses; moreover, aggressiveness is associated with defensiveness in this model, which is a prominent feature of human reactive aggression. Based on the degree of abnormality and severity, we believe that this type of rodent abnormal aggression is the best match for physical forms of human reactive aggression. The instigation model may be a specific subtype of emotional aggression. In this model, aggressive arousal is increased by a physically inaccessible individual, while aggression is directed towards an "innocent" second, which is, however, accessible. Human displaced aggression is in many ways similar.

The correspondence between human instrumental aggression and certain abnormal aggression models was also based on emotional responsiveness. Instrumental aggression is goal directed and is associated with low emotional/autonomic arousal. The first attribute was not evidenced in animals so far; the analogy is based on the second. Nevertheless, excessive and/or abnormal forms of aggression, when

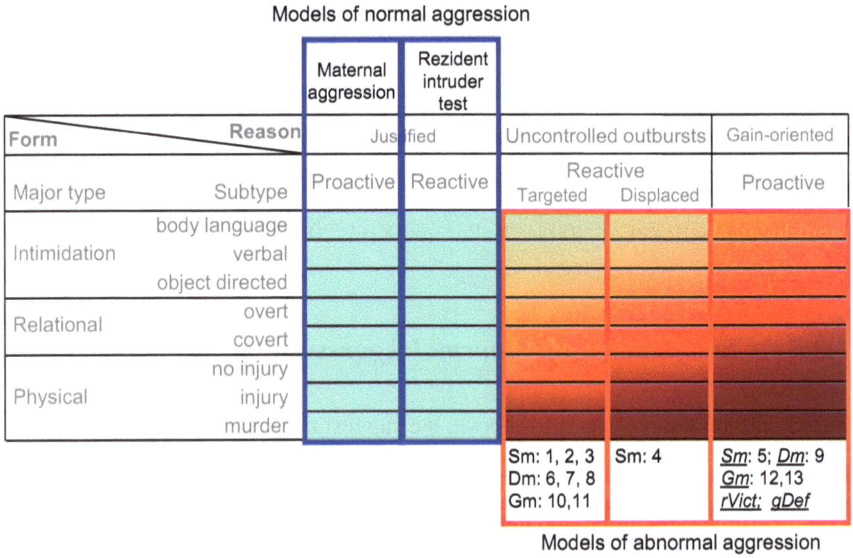

Fig. 1.2 Correspondence between human aggressive behavior and rodent models of normal and abnormal aggression. *Stress models (Sm)*: *1* maternal separation, *2* pubertal stress, *3* post-weaning social isolation, *4* instigation, *5* early subjugation; *Drug models (Dm)*: *6* anabolic steroids (?), *7* amphetamine (?), *8* cocaine (?), *9* alcohol (?); *Genetic models (Gm)*: *10* low anxiety rats, *11* monoamine oxidase A knockout, *12* high-anxiety rats, *13* short attack latency mice; *rVict* repeated victory; *gDef* chronic glucocorticoid deficiency; *question marks in the above list*, emotionality assumed only. For details see Sect. 1.2.3

associated with low emotional responses, may be sufficient to establish the analogy. In addition, the correspondence is supported by similarities between etiological factors and aggression-unrelated behavioral features in many models.

The analogies discussed here are naturally tentative. For example, some of the abnormal aggression models were insufficiently characterized as it regards their emotional background. In these models, emotionality was assumed based on studies unrelated to aggression. For clarity, such models were identified by a question mark in the legends of Fig. 1.2. On the other hand, analogies between instrumental/proactive aggression and hypoarousal-associated models were based primarily on emotional background; the instrumental nature of aggression was not tested in any of the abnormal aggression models. Despite these deficiencies, we believe that the analogies are reasonable and provide testable hypotheses for future studies.

1.2.4 Normal and Abnormal Aggressions in Humans

Human research on aggression differs from that performed in rodents in two major ways. Firstly, "real" aggression (fistfights, weapon use, or murder) cannot be directly (on-site) studied in humans. Therefore, direct observations are limited to substitutes of aggression, e.g., games where subjects "fight" with fictional opponents in rather symbolic ways. In other words, models rather than genuine aggressive acts are studied. Secondly, aggressive humans—unlike animals—have detailed medical and/or criminal records; in addition, people can be asked about their thoughts, feelings, social relationships, etc. Consequently, research on human aggression is extremely rich in studies on aggression histories. By this phrase we mean that aggressiveness is attributed to people based on their past rather than on their current behaviors. For example, violent criminals are considered aggressive even if they readily cooperate with the study personnel and show no aggression during the study.

1.2.4.1 Aggression History-Based Studies

The procedures employed are extremely diverse, but may be grouped into three main approaches: "behavior-oriented," "criminal record-oriented," and "psychopathology-oriented" approaches. All three approaches focus on the history of subjects, not on behaviors momentarily displayed. This is the reason of grouping them under the generic term "aggression history-based studies."

The "behavior-oriented" approach addresses the behavioral (phenomenological) aspects of aggression by employing psychometric instruments. In addition to criminal and psychopathological cases, these instruments are frequently employed to study psychiatrically healthy and non-criminal populations.[5] Some of the questionnaires address all three physical, verbal, and indirect aggressions in structured and systematic ways (*Macarthur Health and Behavior Questionnaire*, Essex et al. 2002) or ask plain questions that allow inferences on these forms of aggression (*Early Adolescent Temperament Measure*, Capaldi and Rothbart 1992; *Composite Aggression Score* by Cullerton-Sen et al. 2008; *Adolescent Aggression Questionnaire*, Orpinas and Frankowski 2001). Other inventories neglect indirect (relational) forms of aggression and focus on verbal and physical aggressions (*Brown-Goodwin Aggression History Scale*, Brown et al. 1979; Coccaro et al. 1997a, b; *Buss-Durkee Hostility Inventory*; Buss and Durkee 1957; *Buss-Perry Aggression Questionnaire*, Buss and Perry 1992; *Agitation/Aggression subscale of the Neuropsychiatric Inventory*; Cummings et al. 1994; *Displaced Aggression Questionnaire*, Denson et al. 2006; *Reactive-Proactive Aggression Questionnaire*, Raine et al. 2006). In addition to aggressive behaviors in general, these inventories may also ask about the form of aggression specifically (e.g., weapon use, reactive-proactive types of aggression, etc.) and quite frequently ask about feelings (angriness or angry rumination, cruelty, defiance, frustration, impulsivity, irritability, jealousy, revengefulness, suspiciousness); moreover, sometimes they investigate

[5] A systematic overview of all the methodologies employed in human aggression research overpassess the scope of this section. We focus here on representative psychiatric instruments.

the ability to recognize emotions (Capaldi and Rothbart 1992). Some of the inventories investigate nonaggressive conduct problems too (lying, cheating) (Essex et al. 2002). Functional aspects of aggression are sometimes also addressed, mostly in the form of negative consequences for social-emotional functioning (Capaldi and Rothbart 1992; Buss and Durkee 1957; Essex et al. 2002).

Based on the short review of a few representative examples, one can conclude that "behavior-oriented" inventories are in line with the phenomenological definitions of aggression; however, indirect (e.g., relational) forms are underrepresented in the inventories overall. Asking about feelings is common. Albeit not directly stated as a rule, feelings are palpably considered causative. Functional aspects—when addressed—focus on negative consequences. This point of view is shared with the one adopted by "psychopathology-oriented" approaches. Addressing goals and potential benefits in terms of competitive advantage is highly uncommon. Even when proactive forms of aggression are expressly addressed (Raine et al. 2006), reasons and successfulness are not investigated in functional terms.

The "criminal record-oriented" approach addresses the "rule-breaking" aspect of aggression. It is practically impossible to overview all the studies that used delinquency as a criterion for aggressive behavior; we will present extremes. In a series of studies on aggression, Popma et al. (2006, 2007a, b) studied subjects who were incriminated with petty offenses, i.e., "wanton destructiveness, vandalism, simple theft, hooliganism, infractions on the firework regulations, and minor forms of aggression." At the other extreme, Raine et al. (1992, 1994, 1997) specifically studied murderers. The rest of the studies are between these two extremes and usually cover a wide range of violent crimes from fistfights to weapon use and murder. Study populations are often heterogeneous as it regards the severity of violent crimes committed.

From the point of view of the theoretical definitions, "criminal record-oriented" studies focus on physical aggression. The study populations quite frequently include antisocial acts that are not aggressive according to the definitions reviewed in the first section (e.g., theft), but rarely if ever cover instances of verbal aggression and even less so indirect (relational) aggression. Thus, "criminal record-oriented" studies may involve harm in economic means, but most often they cover physical harm that is inflicted in a rule-breaking manner. In general, these studies are deficient regarding the evaluation of behavioral and rationale-related aspects. The first derives from the tendency of authors to study together widely different types of violent crimes. From a legal perspective, all the behaviors studied can be categorized as violent crime. From a neurobiological point of view, however, one can at least assume that widely different types of behavior (e.g., hitting and killing) have different underpinnings. As it regards reasons, the "emotional (reactive)-instrumental (proactive)" distinction is neglected quite often. Some people commit violent crimes because they cannot control their emotional outbursts; others become aggressive exactly because they *can* control emotions, which make them capable of using aggression for personal profit. Although this distinction starts to gain terrain (as a sign, the first Reactive-Proactive Aggression Questionnaire was recently developed, Raine et al. 2006), violent criminals are usually studied indiscriminately as it regards the reasons of their behavior.

"Psychopathology-oriented" studies address the dysfunctional aspect of aggression, by concentrating on aggression-related psychiatric disorders usually in terms of the Diagnostic and Statistical Manual of Mental Disorders (DSM). There are six psychopathologies where aggressiveness is expressly mentioned among the symptoms; these were shown in Table 1.1.[6] Some of the symptom lists make reference to all three physical, verbal, and indirect aggressions (conduct disorder and the psychopathic subtype of antisocial personality disorder); others involve verbal but not other types of aggression (oppositional-defiant disorder), yet others entail physical only or both physical and verbal aggressions (antisocial and borderline personality disorders, intermittent explosive disorder). The reasons of aggression may also be deduced from the symptoms. The description of intermittent explosive disorder is an accurate account of reactive aggression, while the description of psychopathy suggests similarity with instrumental aggression. Nevertheless, the reasons of aggressiveness cannot be unequivocally deduced from the symptoms. For example, symptoms indicate that both antisocial and borderline personality disorders are characterized by reactive aggression, while research suggests that the former is mostly associated with proactive while the latter mostly with reactive aggression (de Barros and de Pádua Serafim 2008).

An interesting feature of the "psychopathology-oriented" approach is that subjects are not necessarily aggressive. This is clearly shown by the symptom lists of aggression-related psychopathologies and the number of symptoms required for diagnosis (Table 1.1). For example, antisocial personality disorder can be diagnosed when three of its seven symptoms are met. Yet, only one of the seven symptoms makes explicit reference to aggressiveness, and only two of the remaining six may potentially engender aggressiveness. Thus, an individual can be diagnosed with this disorder in the absence of aggression or aggression-related symptoms. The situation is not different with other disorders; the only exception is intermittent explosive disorder that has only one symptom, and this makes explicit reference to aggressiveness. Yet, this particular symptom is quite often seen in depression (Fava 1998; Painuly et al. 2011); thus, aggression specificity may not be ensured with this disorder either. Taken together, psychiatric constructs focus on dysfunctions. Consequently, aggression-related psychopathologies are only loosely related to the features according to which aggressive behaviors are delimited from other behaviors on theoretical grounds (i.e., subjects may be diagnosed with aggression-related psychopathologies even if they do not show aggressive behaviors).

[6] In the followings, we will refer to mental disorders as described in DSM-IV-TR (American Psychiatric Association 2000).

Table 1.1 The main symptoms of aggression-related psychopathologies based on DSM-IV-TR (American Psychiatric Association 2000)

Conduct disorder (3)	Oppositional-defiant disorder (4)	Antisocial personality disorder (3)	Antisocial, psychopathic type	Borderline personality disorder (5)	Intermittent explosive disorder
1. Bullying, threatening, intimidating 2. Initiating physical fights 3. Weapon use 4. Physical cruelty to people 5. Physical cruelty to animals 6. Steeling while confronting victim 7. Sexual abuse 8. Fire setting with the intention of causing damage 9. Destroying property 10. Breaking into house, building, or car	1. *Losing temper* 2. Arguing with adults 3. Actively defies or refuses to comply with adults' requests or rules 4. *Deliberately annoying people* 5. Blaming others for mistakes or misbehavior 6. *Touchy or easily annoyed by others* 7. *Angry and resentful* 8. Spiteful or vindictive	1. *Failure to conform to social norms with respect to lawful behaviors* 2. Deceitfulness for personal profit or pleasure 3. Impulsivity or failure to plan ahead 4. Irritability and aggressiveness (repeated physical fights or assaults) 5. Reckless disregard for safety of self or others 6. Consistent irresponsibility, in terms of work behavior or financial obligations 7. *Lack of remorse (indifferent to or rationalizing, having hurt, mistreated, or stolen from another)*	1. *Callousness* 2. Aggressive verbally, relationally, or physically to control others 3. Manipulativeness (use of cunning, charm etc., to personal advantage) 4. *Hostility (irritability, rudeness, etc.)* 5. Deceitfulness (fabrication when relating events, etc) 6. Narcissism (vanity, self-centeredness, etc.) 7. Irresponsibility (failure to honor obligations) 8. Recklessness (pursuit of stimulation without regard for consequences) 9. *Impulsivity (acting without a plan or consideration of outcomes)*	1. Frantic fear of abandonment 2. Unstable and intense interpersonal relationships 3. Identity disturbance (unstable self-image) 4. Impulsivity in at least two areas that are potentially self-damaging 5. Suicidal or self-mutilating behavior 6. *Affective instability (episodic dysphoria, irritability, or anxiety)* 7. Chronic feelings of emptiness 8. Inappropriate anger or difficulty controlling anger (e.g., displays of temper, physical fights) 9. Stress-related paranoid ideation or dissociative symptoms	Several discrete episodes of failure to resist aggressive impulses that result in serious assaultive acts or destruction of property

11. Lying to obtain advantage	
12. Stealing without confronting victim	
13. Staying out at night despite parental prohibitions	
14. Truancy from home	
15. Truancy from school	

Note. Symptom descriptions were simplified without changing their essence to fit table. Symptoms explicitly referring to aggression were underlined; those that *may* involve aggressiveness were printed in italic. The number of symptoms required for a diagnosis was indicated between brackets after the name of the disorders

1.2.4.2 "Manifest" Approaches

These models detach from the earlier ones in the sense that they investigate ongoing behavior, i.e., aggressive acts performed during the study. Their prototype is the Taylor Aggression Paradigm (Taylor 1967). Subjects submitted to the procedure are made to believe that they are playing a competitive reaction time task against a living opponent, which is in fact fictional. Subjects are informed that whoever loses a given trial receives punishment from the winner. Punishments are blasts of noise or electric shocks. When subjects lose, they receive punishments of a preset level. Subjects can select the level of punishment delivered to their opponents when they win. Aggressiveness is defined by the level of punishment delivered by subjects overall and by the tendency to escalate punishment when receiving high-level punishments themselves (i.e., when challenged). Another widely used model is the Point Subtraction Aggression Paradigm (Cherek 1992). In this model, players can earn points that are converted into money at the end of the session. They have three options: increasing their earnings, decreasing the earnings of the fictional opponent, or protecting their earnings. Each choice is signaled by pressing buttons. Players are challenged by point subtractions delivered to them by the opponent. The challenge may be responded aggressively by deducting points from opponents in retaliation or nonaggressively by protecting earnings. Aggressiveness is defined by the tendency to retaliate when challenged.

Somewhat more sophisticated modes of aggression are studied by the Ultimatum Game, where subjects receive offers to split money and unfair offer rejections are considered aggressive (Guth et al. 1982); by the Game of Chickens, where subjects can chose cooperation or aggression in a situation where mutual aggression results in money loss, mutual cooperation leads to moderate earnings, while aggression against cooperation is the most profitable tactics (Rapoport and Chammah 1966); by the Survivor Contest, where displaced aggression is measured in terms of money subtraction from innocent victims in response to provocation by others (Reijntjes et al. 2007), etc. An interesting model is the Articulated Thoughts in Simulated Situations Paradigm, which studies verbal aggression elicited by a simulated conflict with the spouse (Eckhardt et al. 1998). Anger arousal by picture viewing and imagined aggression may also be used to study aggression (Rosenzweig et al. 1975; Strenziok et al. 2011); such models are more distant from executive functions than the activity-centered models described above.

The advent of video games offered new ways to study aggression-related executive functions. In addition to being related to real-life aggressiveness (see Anderson et al. 2010 for a recent review), these games offer the chance to study aggression-related decision making and the execution of virtual aggressive acts and to study "virtual cruelty," e.g., by offering the players the opportunity to view the opponent suffering (Fukui et al. 2006; Lotze et al. 2007; Regenbogen et al. 2010; Salminen and Ravaja 2008). Although opponents are known to be fictional in these games (as opposed to the Taylor Aggression Paradigm for instance), the positive relationship between video game and real-life aggression on one side and the correlations between this and other "manifest approach" models on the other suggest that video games activate the mechanisms of aggression.

The procedures reviewed here are in line with the "resource competition" and "harm delivery" criteria of theoretical definitions. Intentionality cannot be questioned either, and the victim is not benefited by any means. Despite the fact that the aggressive response is quite symbolic when compared, e.g., to violent crimes, these models are the best fits for the theoretical definitions. As it regards aggression types, most of the procedures imply retaliation, suggesting that they model reactive (emotional) aggression. This is especially blatant in models where aggressiveness leads to monetary loss, i.e., the subjects are motivated by emotions rather than by interests. The only exceptions may be aggressive video games that have a strong instrumental/proactive component: players deliberately engage in obtaining advantage (entertainment) by inflicting virtual harm. Committing antisocial acts "for personal profit or *pleasure*" is expressly mentioned in the symptom list of antisocial personality disorder; thus "entertainment" may be a valid instrumental goal of aggression, suggesting that the motivation is realistic. Harm infliction is virtual; noteworthy, however, the screen images of harm are highly naturalistic, which may close up the gap between real and virtual aggression.

1.2.5 Classifications of Operational Approaches in Humans

The correspondence between theoretical definitions on one side and "behavioral," "criminal," and "manifest" approaches on the other was visualized in the upper panel of Fig. 1.3.

Some of the "behavioral approaches" demarcate the phenomenology of aggression. These cover all main forms of aggression (intimidation, relational, and physical) and to a certain extent their subtypes. The level of elaboration and the behavioral forms that are covered are inventory specific, but overall, neither is particularly concerned with the reasons of aggression or with their abnormality. Other "behavioral approaches" are more specific regarding either the form or the reason of aggression.[7] The Articulated Thoughts in Simulated Situations Paradigm, for instance, studies reactive verbal aggression; the Reactive-Proactive Questionnaire aims at differentiating two basic forms of rule-breaking aggression, while the Displaced Aggression Questionnaire specifically focuses on the various types of displaced aggression.

Criminal record-oriented approaches by contrast focus on the rule-breaking feature of aggressiveness, primarily on physical aggression. Some of these approaches address severely violent crime, while others consider destructiveness that can be considered object-oriented displaced aggression with a criminal dimension.

---

[7] To avoid an overcrowded figure, we make use of examples.

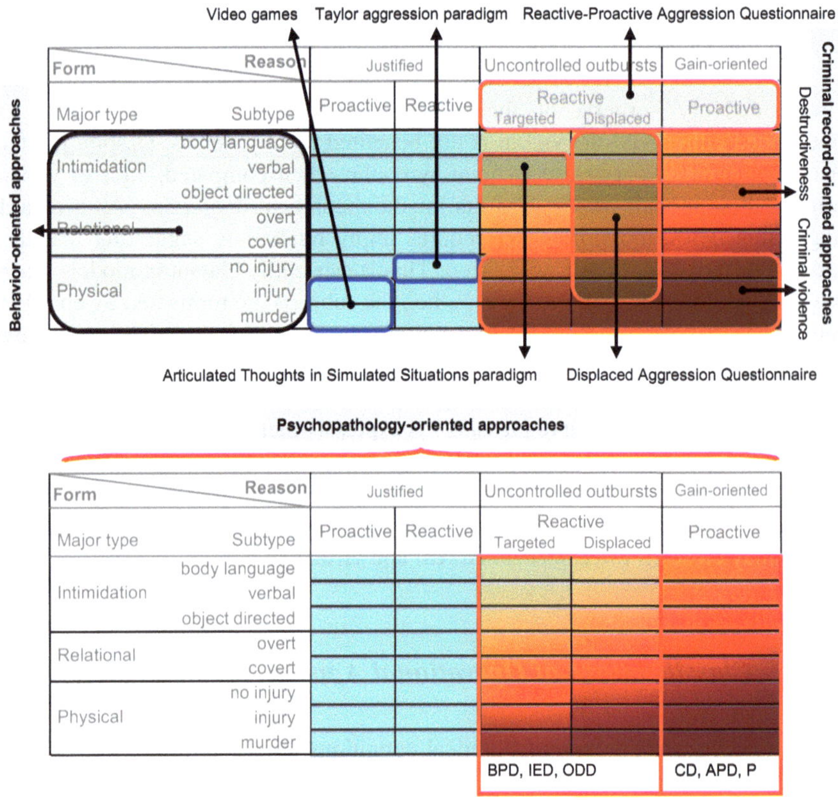

Fig. 1.3 The theoretical system of aggressive behaviors and operational approaches in humans. *Rounded squares*, aggression types covered by the approaches indicated by *arrows*; *black*, applicable to both justified and abnormal aggression; *blue*, justified aggression; *red*, abnormal aggression; *APD* antisocial personality disorder; *BPD* borderline personality disorder; *CD* conduct disorder; *IED* intermittent explosive disorder; *ODD* oppositional defiant disorder; *P* psychopathy. Detailed in Sect. 1.2.5

Finally, aggressiveness studied by "manifest" models is usually physical in nature; this can be virtual like in the case of video games or real when subjects deliver punishments in the form of electric shocks or blasts of noise. An important feature of "manifest" approaches is that they model normal forms of aggression, because (1) the harm inflicted is either minor (shocks are mild, the value of the money subtracted is small) or virtual (video games) and (2) subjects do not show rule-breaking behavior. Rules are set by the experimenters; moreover, the means of aggression are also provided by the study personnel. Subjects do not break rules; moreover, they follow the instructions received. The upper panel of Fig. 1.3 includes aggressive video games and the Taylor Aggression Paradigm as models of proactive and reactive normal aggression, respectively.

Psychopathology-oriented approaches on their turn investigate a mixture of features (Fig. 1.3, lower panel). Both the forms and the reasons of aggression are considered. Based on symptoms, one can hypothesize that borderline personality, oppositional-defiant, and intermittent explosive disorders are associated with reactive aggression ("uncontrolled outbursts"), while conduct disorder, antisocial personality disorder, and psychopathy are associated with proactive aggression. This categorization, however, should be used with caution. The diagnosis per se is not sufficient to make assumptions on the forms, reasons, or severity of aggression; moreover, they are not sufficient to assume the mere presence of aggression with the exception of intermittent explosive disorder.

1.3 Definitions, Approaches, and the Validity of Conclusions

According to theoretic definitions, the term "aggression" covers a wide variety of behaviors that range from threatening body postures to instrumental murder in humans and from offensive threats to biting attacks in rodents. A comprehensive circumscription of this behavior on the theoretical level demanded three sections, and its graphical representation required a three-dimensional space. Operational approaches on their turn do not and cannot cover "aggression" as a whole; on the contrary: they quantify particular aspects of this behavior. Moreover, these approaches are used to study aggression on widely different time scales: some focus on a history of aggression, while others tackle ongoing behaviors. The diversity of aggressive acts is one of their intrinsic features, while the diversity of operational approaches can be considered a reflection of this inherent diversity. As such, the multitude of methodologies should be perceived as an asset; they allow a multidimensional approach to a multifaceted phenomenon. These diversities, however, raise an important question: how can one generalize conclusions based on one particular methodology to aggression as a whole?

In his attempt to evaluate the validity of conclusions, Aristotle (around 330 BC) differentiated universal and particular premises. The former can be characterized by the word "all" while the latter by the word "some." One of the basic ideas of Aristotle's syllogistic method is that the characteristics of the premises define the characteristics of conclusions; no universal conclusion can be drawn if one of the premises is particular. This basic idea—when translated into the practice of aggression research—means that universal conclusions on aggressive behavior can be reached only if the premises are universal. Neither the theoretical definitions nor operational approaches seem to provide a basis for formulating universal conclusions as it regards aggression.

Even intensional definitions require specifications regarding the meaning of the word "harm," which forecasts even larger diversity when extensions are considered. According to current concepts, various forms of aggression can be classified along at least three dimensions: behavioral form, reason, and abnormality (see Fig. 1.1). One can reasonably doubt that the mechanisms underlying threatening body language and those underlying murder for profit are similar. Similar diversities can be observed in the case of operational approaches (Fig. 1.3). Killing virtual opponents for pleasure (video games) is likely controlled by other mechanisms than those controlling belligerent provocations that are articulated during anger arousal in the modified version of the Articulated Thoughts in Simulated Situations Paradigm. The comparison of human and animal models and, moreover, the comparison of rodent aggression models themselves raise further questions regarding the validity area of conclusions (Fig. 1.2). Similarly, mechanisms that increase the incidence of being engaged in aggressive conflicts over the life span (or over prolonged periods) are likely different from those, which are acutely activated during aggressive conflicts. These concerns regarding the generalizability of conclusions can easily be substantiated by experimental findings. The brain control of reactive and proactive aggression was suggested to differ substantially (Blair 2001; Lopez et al. 2004); it was repeatedly suggested that these two forms should be studied separately, and the Reactive-Proactive Aggression Questionnaire was expressly developed to reveal the behavioral differences between them (Raine et al. 2006). In the same vein, a history of aggressive delinquency was associated with decreased orbitofrontal functions (Birbaumer et al. 2005; Boccardi et al. 2011; Raine et al. 1992), while the very same area was overactivated when subjects imagined aggression, played aggressive video games, or delivered physical punishments in competitive situations (King et al. 2006; New et al. 2009; Strenziok et al. 2011; Veit et al. 2010). Aggression type-specific mechanisms were shown in rodent studies as well; e.g., the brain areas activated by fights in the chronic glucocorticoid deficiency model included regions that were not activated in normal resident-intruder conflicts (Halasz et al. 2002a).

These considerations suggest that the mechanisms of aggression control cannot be studied in general. In other words, syllogisms of the following type:

Major premise: Aggression is typical in conduct disorder,
Minor premise: Conduct disorder is associated with low plasma glucocorticoids,
Conclusion: Aggression is associated with low plasma glucocorticoids,

should be replaced with syllogisms of the following type:

Aggression is typical in conduct disorder,
Conduct disorder is associated with low plasma glucocorticoids,
Some forms of aggression are associated with low plasma glucocorticoids,

because conduct disorder is a particular—moreover, rather specific—operational approach to the broader phenomenon "aggression."

The considerations outlined above strongly suggest that the quest for *global* mechanisms of aggression control is an elusive goal. Findings obtained with one paradigm cannot be generalized to aggression as a whole; such an approach blurs rather than clarifies the general picture. This does not necessarily mean that each form of aggression has its specific background, and each extension of theoretical definitions and all the operational approaches have a differential neural underpinning. It seems likely that the number of mechanisms is smaller than the number of tests. One can hypothesize that one and the same mechanism subserves several types of aggressive behaviors, the exact form of which may be determined by the context. The generalizability of conclusions, however, should be studied and carefully evaluated rather than taken as granted. This is the general principle adopted in the next sections where the endocrine and neural underpinnings of various forms of normal and abnormal aggressions are discussed.

Chapter 2
Hormonal Determinants

2.1 Hormones and Brain Function

The presentation of hormonal effects may be surprising in a work that focuses on the neural control of aggression. Hormones affect the organism in multiple ways, but none of these involves the direct activation of neurons or neuronal circuits. As such, they are not able to directly control the execution of behaviors. However, hormones have a large impact on how the neural system works. The neural mechanisms discussed in the next chapter are under hormonal control; moreover, dysfunctional aggression likely results from neural changes that are prompted by changes in hormone secretion patterns. As such hormone secretion patterns can be viewed as explanatory mechanisms for the neural control of aggression.

It is important to note that hormones affect brain function in an anatomically poorly specific manner: hormonal effects are exhibited wherever their receptors are expressed. Thus, hormonal effects are unspecific, which raises an important question: how do the nonspecific neural effects of hormones translate into specific behavioral roles? This question was frequently asked when I lectured on the relationship between glucocorticoids and aggression. The response to this question is that the specific role of hormones in behavioral control is determined not by the behavior specificity of hormonal effects but by the specific relationship between behaviors and brain activation patterns.

On a neural-mechanistic level, the previous statement can be explained in the following way. Any behavioral act is controlled by a specific set of neurons that form a network. If the properties of neurons are altered by hormonal influences, the functioning of the network is also altered, and consequently, the behavior controlled by the network will be changed. The effects of hormones are naturally not restricted to those neurons that control a particular behavior. However, effects on networks unrelated to the context (i.e., are not "working" in that particular moment) remain "silent" because they do not contribute to the execution of the behavioral act. According to this model, the specificity of hormonal effects derives from the context (Fig. 2.1). This mechanistic model can be complemented by situation-

J. Haller, *Neurobiological Bases of Abnormal Aggression and Violent Behaviour*,
DOI 10.1007/978-3-7091-1268-7_2, © Springer-Verlag Wien 2014

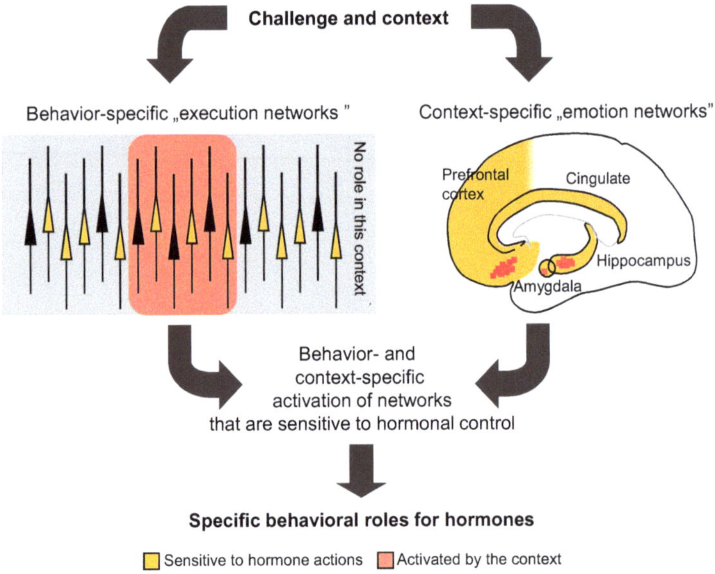

Fig. 2.1 *A theoretical model of the behavioral specificity of neurally non-specific hormonal actions*. The main point of this figure is that the behavioral specificity of hormonal actions is not defined by their neural specificity but by the behavior-specificity of brain activation patterns. *Upper left* schematic representation of a layer of cortical pyramidal neurons. While hormones affect the properties of many neurons, the consequences of this are manifested only when the neurons in question contribute to the elaboration of the actual behavioral response. *Upper right* schematic representation of the main limbic areas that control emotional responses. Emotions are associated with a specific pattern of limbic activation as shown by brain imaging studies. Again, hormones affect emotional responses specifically by virtue of the specificity of such activation patterns

specific effects on emotional responses. In contrast to neural networks controlling movements (e.g., behavioral acts), emotions are controlled by extended brain regions commonly known as the limbic system, which is subject to hormonal effects as a whole. Emotions, however, do not involve the activation of the whole limbic system. On the contrary, particular emotions are associated with particular activation patterns as shown by brain imaging studies and amply detailed later. Consequently, hormones may indeed have overall effects on the limbic system, but only those effects become manifest under particular situations, which are related to the context.

We illustrate these by taking testosterone as an example. This hormone affects brain mechanisms involved in sexual behaviors but also those that are involved in aggression. While both mechanisms are promoted by testosterone, its effects on sexual behavior cannot be observed when the subject is challenged aggressively. The behavioral effects of testosterone are made specific by the context.

We discuss below the roles of two hormones: testosterone and glucocorticoids. These are not the only ones that affect aggression but seem more important than others for several reasons: (1) both hormones affect neural development by genomic actions. Aggression-related psychopathologies seem to be developmental disorders; therefore, studying mechanisms that influence neuronal development are revealing on how aggression-bound personalities arise; (2) both are highly sensitive to environmental influences and affect gene expression profiles throughout the lifespan. By these processes they continuously shape the function of the neural system and have an impact on aggression-related traits; (3) both have rapid, non-genomic effects, by which they are able to control ongoing behaviors.

In addition, they continue to be hot topics in aggression research, despite the fact that their roles were discovered rather early. While many studies attest their important role, contrasting findings are also frequent. We will argue below for the idea that contradictions are not due to experimental errors or misjudgment, but derive from the complexity of hormonal effects and the complexity of aggressive behavior. We will show that the role of hormones in aggression cannot be described in simple terms like "promote" or "inhibit." Their effects depend on the constellation of several factors among which the context is likely the most important one. Both hormones are deeply involved in the control of aggression but not in all types of aggression, not in every context, and not linearly.

2.2 Testosterone

The dependence of animal aggression on testosterone secretion was scientifically recognized in the mid-nineteenth century (see below) and was amply confirmed by animal research ever since. The general idea is that the role of testosterone in aggression derives from its role in sexual behavior, as the ability to monopolize resources is a crucial factor of reproductive success, while access to resources depends to a large extent on aggressiveness. Testosterone is believed to be a common controlling factor of both. By contrast, the role of testosterone is considered uncertain in humans, where aggression has a minor role in sexual competition. Findings support neither assumption. Animals can be aggressive without testosterone, and evidence linking testosterone to human aggression is more robust than generally thought. The findings reviewed here, together with the theories concerning the role of testosterone, strongly support the notion that testosterone is an important mechanism of aggression control and significantly contributes to the emergence of abnormal forms of aggression. It will also be shown that aggression is not controlled by testosterone alone, and its role is embedded in a wider array of controlling mechanisms. We will argue that the role of testosterone cannot be understood without incorporating it into the context of other controlling mechanisms.

2.2.1 Mechanisms of Action

Similar to other steroid hormones, testosterone exerts its effects by modulating gene expression. Briefly, testosterone enters the cytoplasm where it remains intact or is metabolized to dihydrotestosterone or estradiol, which binds to androgen or estrogen receptors, respectively, and affects the transcription of genes that express androgen or estrogen response elements. Such hormone receptors and hormone response elements are nearly ubiquitously expressed, which explains the multitude of testosterone effects that range from the control of sperm production to aggression but at the same time makes it difficult to pin down the exact mechanism by which one particular phenomenon (e.g., aggression) is regulated by genomic effects.

In addition to the widely recognized role of testosterone in gene expression, recent findings suggest that this hormone may affect neuronal function and behavior by non-genomic mechanisms. In contrast to genomic effects, non-genomic ones are exerted rapidly and may affect ongoing behavior. Unfortunately, the non-genomic effects of testosterone are poorly known at present, and to our knowledge, there are no convincing reports on their implication in aggression control. By contrast, such non-genomic effects were implicated in the regulation of vascular and sexual functions, sexual behavior, social communication, and cognition (Balthazart et al. 2009; Cho et al. 2003; Cornil and Charlier 2010; Heinlein and Chang 2002; Walker 2003; Nyby 2008). Non-genomic behavioral effects are mediated by effects on membrane fluidity, interactions with intracellular calcium regulatory mechanisms, the activation of second messenger pathways (e.g., the mitogen-activated protein kinase), and modulatory actions on membrane receptors ($GABA_A$, NMDA, glycine, nicotinic, and vasopressin V1a receptors) (Foradori et al. 2008; Gabor et al. 2012). In the absence of direct evidence on aggression, all effects will be discussed as being induced by genomic mechanisms, while the involvement of non-genomic ones seems quite likely.

It is important to note here that the activation of mechanisms influenced by testosterone may not entirely depend on its plasma levels. A large share of this hormone is sequestrated by sex hormone-binding globulin, and only free testosterone is ready to enter the cells. Binding and release from this globulin is a dynamic process influenced by many factors, not lastly by its amount. While the issue was not investigated systematically, a few studies indicate that the plasma levels of sex hormone-binding globulin free testosterone have a role in determining the degree to which testosterone can influence aggressive behavior in both animals and humans (Aluja and García 2007; Brambilla et al. 2001; Pajer et al. 2006; Witte et al. 2009). Such interactions may be especially relevant to short-term (possibly non-genomic) effects of testosterone, where the momentary availability of this hormone may have a large impact. Other modifiers of testosterone action are the enzymes that convert it into active metabolites, e.g., dihydrotestosterone and estrogen. The expression level and activity of 5α reductase and aromatase greatly influences the ability of testosterone to influence aggressiveness (Bethea et al. 2013; Compaan et al. 1994a, b; Schlinger and Callard 1990; Soma et al. 2003; Wu et al. 2009). Finally, the expression of receptors that mediate the effects of testosterone and its active metabolites also has an impact on testosterone effects (Bergeon Burns et al. 2013; Canoine et al. 2007; Scordalakes and Rissman 2004; Trainor et al. 2006). Noteworthy, testosterone effects are often influenced by these modifiers in a brain area-specific manner.

Taken together, these findings suggest that evaluating the effects of testosterone is not an easy task. For example, reduced plasma levels of the sex hormone-binding globulin, brain area-specific enhancement of aromatization levels, or receptor expressions may make testosterone efficient even if plasma levels are low. As such, negative findings may be misleading when only testosterone levels are measured.

2.2.2 Developmental Effects on Aggression in Animals and Humans

2.2.2.1 Normal Aggression

The first effects of testosterone on animal aggression are exerted in a critical perinatal period. At this early age, testosterone acts as an organizing principle that makes the brain sensitive to its own effects that are exerted much later, beginning with the pubertal period (Fig. 2.2). This "organizing" role of testosterone was first recognized in conjunction with sexual behavior. It was stated already in the 1960s that "Sex hormones of the embryo affect ultimate sexual behavior in two ways: indirectly by their control of morphogenesis of the genitalia and, in some species at least, directly by their neural-organizer effect" (Money 1965). It was soon discovered that testosterone has similar effects on aggressive behavior: neonatally administered testosterone increased, while neonatal castration decreased adult aggression in mice (Bronson and Desjardins 1968, 1969; Edwards 1969; Peters et al. 1972). This increase in aggression was found to be conditional; instead of leading to an overall increase in aggressiveness, neonatal testosterone in fact sensitizes the neural system to its later effects: the aggression-heightening effects of testosterone in adulthood are facilitated by perinatal exposures to testosterone (Hutchison 1974; Klein and Simon 1991). The early effects of testosterone are so strong that it can induce male-like aggressive behavior in female rodents, provided that these are treated with testosterone in adulthood (Gandelman et al. 1979). Moreover, intrauterine position, e.g., the sex of neighboring embryos, also has a role in establishing adult aggressiveness by virtue of the diffusion of testosterone through the amniotic fluid (Quadagno et al. 1987; Ryan and Vandenbergh 2002). The early organizing effects of testosterone with regard to its later behavioral effects are bound to a specific time window of animal development; this time window is species dependent and short; it encompasses a few hours or days shortly

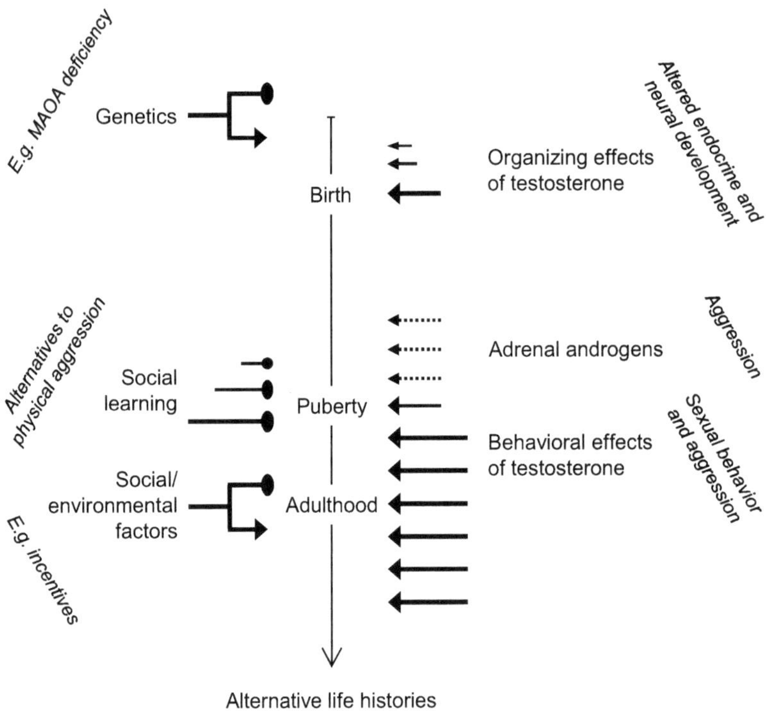

Fig. 2.2 *The two waves of testosterone actions*. The organizing effects of testosterone around birth includes changes in the development of the endocrine system (testosterone production, aromatization, androgen/estrogen receptor expression) of brain structure (size of various brain areas) and brain function (altered neurotransmission). These determine the way in which testosterone affects behavior in adulthood (including effects on endocrine variables). Around puberty, adrenal androgens (converted into testosterone and estrogens in the brain) may also mediate the effects of early testosterone exposure. The *left-hand* side of the graph illustrates the non-deterministic nature of testosterone effects, that may be augmented (*arrows*) or inhibited (*oval pointed heads*) by other factors. Such factors are only exemplified

before, during, or after birth (Dixson 1993; Klein and Simon 1991; Motelica-Heino et al. 1993).

These findings demonstrate that there is critical time period in the early development of animals, when testosterone "transforms" the brain such that it becomes sensitive to its own aggression-promoting effects in adulthood. Similar changes were seen in humans, where the prenatal period is also considered crucial for the sexual differentiation of brain and behavior (Warne et al. 1977; Rubin et al. 1981). This assumption is supported by three lines of evidence: (1) congenital adrenal hyperplasia, a condition that is associated with high testosterone levels in early periods of life, results in male-like aggressive behavior in females (Berenbaum et al. 2000; Hines et al. 2002); (2) prenatal exposure to androgens via treatments received by mothers increased aggressiveness in both male and female offsprings (Reinisch 1981); and (3) the length ratio of the second and fourth fingers (a somatic marker of prenatal testosterone exposure called 2D:4D digit ratio) associates negatively with aggressiveness, suggesting that larger testosterone exposure during embryonic life (when digit length is determined) is associated with higher aggression during adulthood (Butovskaya et al. 2013; Hampson et al. 2008; van der Meij et al. 2012). Moreover, girls of opposite-sex twins show a more masculine pattern of aggression than girls of same-sex twins, which may be considered a human analog of the intrauterine position effect seen in animals (Cohen-Bendahan et al. 2005).

Taken together, the findings reviewed above suggest that the developmental effects of testosterone are rather similar in animals and humans. Contradictory findings also exist. Voracek and Stieger (2009), for instance, found no association between 2D:4D digit ratio and aggression, and the first author of the above paper questioned the validity of the findings of van der Meij et al. (2012) based on data-analytic and statistical power issues (Voracek 2013). Such contradictions may be solved by taking aggression types and genetic constitution into consideration. Millet and Dewitte (2009) for instance found that 2D:4D digit ratio predicted aggression in the dictator game (a variant of the ultimatum game presented in Chap. 1) only when an aggression cue was present; in neutral situations, the low (more masculine-like) digit ratio predicted prosocial behavior. In line with this, Benderlioglu and Nelson (2004) found that a low 2D:4D digit ratio predicted aggressiveness in females under high provocation only. These authors deduced aggressiveness from the follow-up letters of subjects solicited to give donations for a fictitious charity organization. Under these conditions, digit ratios predict aggressiveness in women but not in men. In a study investigating trait aggression, the expected correlation with digit ratios was found in males but not in females (Bailey and Hurd 2005). Congenital adrenal hyperplasia by contrast increases aggression in females but not males (Pasterski et al. 2007). In another study, trait aggressiveness was predicted by digit ratios in conjunction with dopamine D4 receptor polymorphism (Butovskaya et al. 2012). Such findings may be considered contradictory, but in our view the differences reflect genetic background-, gender-, and aggression type-related differences in the involvement of testosterone.

2.2.2.2 Abnormal Aggression

Early treatment with testosterone induces male-like aggression patterns in female rodents, which may be considered abnormal by virtue of the second criterion of abnormal aggression (see Chap. 1, Sect. 1.2.2). Particularly, the criterion "disregard of species-specific rules" applies here and renders male-like behavior abnormal in females. As a similar phenomenon is noticed in congenital adrenal hyperplasia, the first criterion of abnormal aggression may also be met (mimicking an etiological factor of human abnormal aggression). Nevertheless, aggression is not a major symptom of this condition, which cannot be considered an aggression-related disorder. Moreover, aggressiveness is increased by this condition in females but not in males, a difference likely related to the reversal of sex roles. In contrast to the human case, early testosterone administration increases adult aggressiveness in rodents belonging to both sexes. Therefore, the resemblance between the human condition and the model is only vague. Beyond this equivocal similarity, there is no information on the nature of aggression elicited by early testosterone manipulations. This does not mean that such treatments result in normal forms of aggression in rodents; it only means that the issue was not investigated so far.

The relationship between early androgen hyperfunction and abnormal aggression was almost as poorly studied in humans as in rodents. There are only two studies suggesting a link. Liu et al. (2012) showed that early androgen exposure (as assessed by digit ratios) increased externalizing behavior in boys but not in girls. A similar association was recently found between digit ratios and intimate partner violence (Romero-Martínez et al. 2013). The only "flaw in the scheme" is that neither externalizing behavior nor intimate partner violence is an aggression-related psychopathology. Yet, a 24-year-long study of externalizing symptoms demonstrated that these often develop into, while intimate partner violence is frequently associated with antisocial behaviors; moreover, psychopathy (Fowler and Westen 2011; Reef et al. 2011). This suggests that early androgen hyperfunction may lead to abnormal human aggression, but the issue remains poorly studied.

2.2.2.3 Mechanisms of Developmental Effects

Perinatal testosterone hyperfunction (1) increases energy metabolism (Quadagno et al. 1977; Hill et al. 2012); (2) increases adult testosterone levels, the aromatization of testosterone to estrogens, and estrogen receptor expression (Ryan and Vandenbergh 2002; Schlinger and Callard 1989; Vaillancourt et al. 2012); (3) stimulates cell growth and differentiation in brain areas related to male-typical behaviors and aggression, e.g., the hypothalamic sexually dimorphic nucleus and its human equivalent, the 3rd interstitial nucleus of the anterior hypothalamus, the ventromedial hypothalamus, and amygdala (Hines 2010; Ryan and Vandenbergh 2002); and (4) upregulates the noradrenergic and dopaminergic innervation of the prefrontal cortex and downregulates the serotonergic innervation of the same area (Butovskaya et al. 2012; Dominguez et al. 2003; Stewart and Rajabi 1994). In addition it masculinizes the serotonergic innervation of the medial preoptic area in female rats (Simerly et al. 1985). Early testosterone treatments also affect the neuronal connectivity (Hines 2010). Importantly, findings in rodents and humans were highly compatible. Some of these testosterone-associated changes may be epiphenomena, but others may be causally linked to changes in aggressiveness.

2.2.2.4 Overall Evaluation of Developmental Effects

Summary of Findings Findings in rodents clearly show that testosterone triggers the development of a testosterone-responsive regulatory system for aggressive behavior within a critical period of early life. The delayed behavioral effects of testosterone are likely mediated by its long-term effects on testosterone production, aromatization, and receptor expression; on the development of aggression-related brain areas; and on the development of noradrenergic, dopaminergic, and serotonergic neurotransmission. Findings in humans are highly compatible with those obtained in animals. The impact of early testosterone on abnormal aggression was not studied in animal models, but findings obtained in humans suggest its involvement in aggression-related psychopathologies.

Relevance Early development is crucial for adult functioning; in the same vein, the early effects of testosterone on the development of aggression-related brain mechanisms may be crucial for aggressiveness expressed in adulthood. These functional and anatomical consequences of early testosterone exposure suggest that even normal adult levels of testosterone confer the individual a high predilection towards aggressiveness; moreover, early development may predispose individuals to abnormal forms of aggression if plasma testosterone levels increase. We note that while the available information can be ordered in a meaningful way, developmental studies are scarce, and many issues await clarification.

2.2.3 Postpubertal Effects in Animals: Normal and Abnormal Aggressions

2.2.3.1 Testosterone in Animals: Normal Aggression

Albeit castration was used for millennia to pacify domesticated male animals that are otherwise difficult to handle, the first scientific study attesting the role of the testes in aggression was performed by Arnold Berthold in 1847: he showed that castration decreased, while testis transplantation restored aggressiveness in domesticated roosters (cited based on Soma 2006). The particular role of testosterone was demonstrated considerably later by Seward (1945) and Beeman (1947) who showed that puberty-associated increases in testosterone are associated with the emergence of aggressiveness, while aggression suppressed by castration is restored by testosterone, respectively. These basic findings were amply replicated by subsequent research, which revealed further aspects of testosterone action. Seasonal variations in testosterone production covariate with aggressiveness (Butterfield and Crook 1968); current levels of circulating testosterone are in tight correlation with the propensity to behave aggressively (Rose et al. 1971); testosterone injected after the winning or the losing of aggressive encounters increases aggressiveness and decreases submissiveness, respectively, in subsequent encounters; and finally, social rank correlates positively with plasma testosterone in social groups (Lucion et al. 1996; Solomon et al. 2009; Trainor et al. 2004). This set of congruent findings resulted to a strengthened belief that there is a clear-cut causal relationship between testosterone production and aggression in animals—as opposed to humans, where this interaction is much less certain.

In fact, however, the testosterone-aggression link may be perceived as being just as volatile in animals as in humans. Albeit castration reduced aggressiveness in most studies published so far, there still exist carefully performed studies where this effect was not seen (*California mice*: Trainor and Marler 2001; *lizards*: Moore and Marler 1987; *marmosets*: Dixson 1993; *prairie voles*: Demas et al. 1999; *rock hyrax*: Manharth and Harris-Gerber 2002; *Siamese fighting fish*: Weiss and Coughlin 1979; *song sparrows*: Soma et al. 2000; *Siberian hamster*: Jasnow et al. 2000; *wood rats*: Caldwell et al. 1984); moreover, castration increased aggression in some species under certain circumstances (*in laboratory mice, castration increases aggression towards lactating females*: Haug et al. 1984; *in mandarin voles, castration leads to an overall increase in aggressiveness*: He et al. 2012). Thus, surprising effects of castration were seen in all major vertebrate classes and in a variety of taxa within mammals, including rodents. Interestingly, testosterone production was not totally disrupted from aggression control in these species; e.g., winning-related increases in testosterone facilitated aggression in subsequent encounters in California mice, despite the fact that castration did not reduce aggression in this species (Fuxjager et al. 2011).

Similarly, strange effects were noticed in studies investigating the correlation of aggressiveness with plasma testosterone levels, or the effects of testosterone administration. For example, basal plasma levels of testosterone did not show correlation with aggressiveness in Siberian dwarf hamsters (Castro and Matt 1997), wild-type rats maintained in the laboratory (Everts et al. 1997), the red jungle fowl studied in the laboratory (Johnsen and Zuk 1995), and in laboratory rats in the visible burrow system (Blanchard et al. 1995). Again, the relationship between testosterone and aggression was not totally disrupted in these models. In the study by Johnsen and Zuk (1995), basal levels did not, while acute increases in testosterone did correlate with aggressiveness; in the study by Blanchard et al. (1995), submission was not correlated with low testosterone when group members living in the visible burrow system were preselected for high aggressiveness, but the expected relationship was found when group members were not preselected for aggressiveness. As it regards testosterone administrations, these did not affect aggression in some studies (*ground squirrels, postmating period*: Millesi et al. 2002; *male lambs, familiar environment*: Ruiz-de-la-torre and Manteca 1999; *European stonechats, nonbreeding season*: Canoine and Gwinner 2002); moreover, testosterone *decreased* aggression in Siberian hamsters (Jasnow et al. 2000). As shown above for other models, aggression was not totally disrupted from regulatory control by testosterone, as treatments ineffective in the nonbreeding season or in familiar social environments became effective in the breeding season and in unfamiliar social environments (Canoine and Gwinner 2002; Millesi et al. 2002; Ruiz-de-la-torre and Manteca 1999).

The studies briefly reviewed above (and a series of others not cited here) show that there is no clear cause–effect relationship between testosterone production and aggressiveness. Even more strikingly, strange effects were noted in species where testosterone clearly controls aggression. For example, castration dramatically decreased, but did not *abolish* biting attacks in castrated mice; in addition, the surgery did not affect the frequency and duration of offensive threats (Clipperton-Allen et al. 2011). In male laboratory rats, Albert et al. (1986) reported that the presence of females increased aggression in both intact and castrated males. Thus, the absence of testosterone did not prevent males from being aggressive, not even in species where the relationship between aggressiveness and testosterone was believed to be clear-cut.

2.2.3.2 Is Testosterone "Responsible" for Aggression in Animals?

Asking the same question in a more general way: does a considerable body of contradictory findings question the role of testosterone in aggression? From a simplistic view it certainly does: the assumption that testosterone controls aggression is negated by these findings. However, testosterone does not control aggression: it affects the properties of neuronal circuits, which are under multiple influences. In other words, the neural and behavioral responses to social challenges are not defined by testosterone, but by the overall outcome of the multitude of concurrent influences. Interactions of this kind were systematically mapped in a few species, and several theories were advanced to explain the discrepancies reviewed above (Fig. 2.3).

The first theory of this kind, the "challenge hypothesis" put forward by Wingfield et al. (1990), made a distinction between breeding season-related changes in testosterone production and those elicited by male–male interactions. The theory was based on data obtained in monogamous birds. It was hypothesized that maintaining high testosterone levels for the whole breeding season is detrimental in terms of both increased metabolic costs resulting from enhanced behavioral activity and reduced parental care, a "by-product" of enhanced testosterone production. Metabolic and parental trade-offs are solved by limiting testosterone production to the early phases of sexual competition and periods of social challenge in the parental period. Thus, the disruption of the testosterone-aggression link is ostensible; in fact, testosterone does control aggression, but its production is controlled by "needs." This influential hypothesis was supported by a number of field and laboratory observations in birds and mammals (Cavigelli and Pereira 2000; Ferree et al. 2004; Muller and Wrangham 2004; Ros et al. 2002) and was exploited to explain the testosterone-aggression link in humans (Archer 2006).

The second theory addressing the disruption of the testosterone-aggression link was also based on data obtained in birds. This theory stipulates that nonbreeding season-related aggression (which is rather high in many species) is not controlled by testosterone but by adrenal steroids, e.g., dehydroepiandrosterone (Soma 2006; Soma et al. 2008). The mediator of this mechanism is melatonin, which is secreted in high amounts during the winter. Under the influence of this hormone, dehydroepiandrosterone is converted to estrogen (one of the active metabolites of testosterone) in the brain. Albeit based on bird data in the first place, this theory also assumes validity for other vertebrates (Soma et al. 2008). Albeit humans are no seasonal breeders, dehydroepiandrosterone blood levels correlate with certain

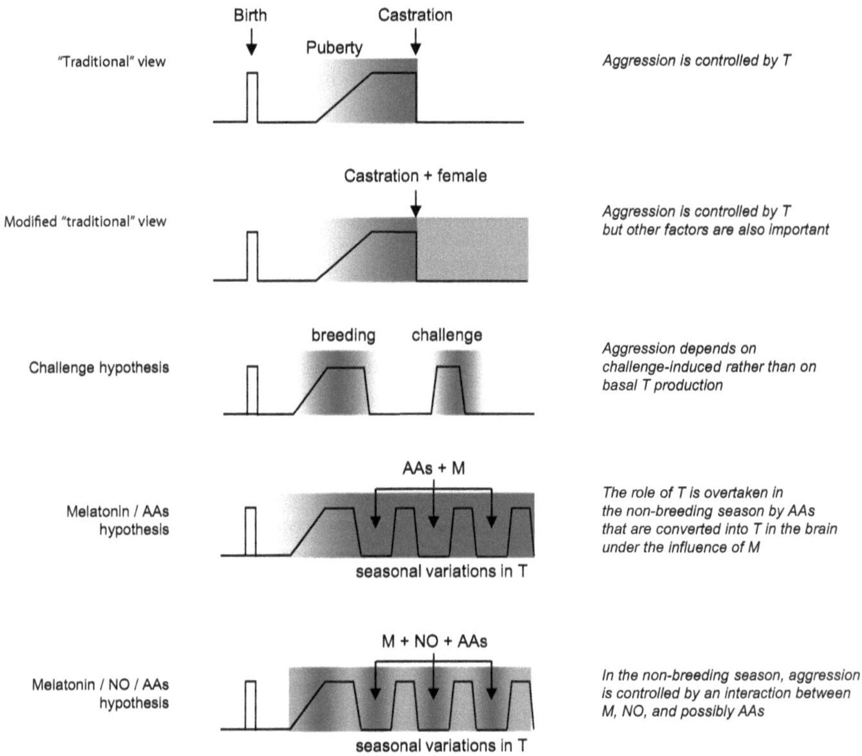

Fig. 2.3 *Hypotheses on the role of testosterone and alternative mechanisms in animal aggression. zigzag lines*, testosterone levels; *gray colour and its intensity*, presence of aggression and its itensity; Letters indicate alternative mechanisms. *AA*, adrenal androgens; *M*, melatonin; *NO*, nitric oxyde. The main point of this figure is that the testosterone/aggression link is not absolute in animals; powerful alternative mechanisms do exist

forms of aggressiveness, suggesting that the mechanisms operating in animals are present in humans (Buydens-Branchey and Branchey 2004; Dmitrieva et al. 2001; van Goozen et al. 1998a, b, 2000a, b). Noteworthy, however, dehydroepiandrosterone does not affect aggression in a species that is critical for this hypothesis—the Siberian hamster, a seasonal breeder which shows high levels of aggression in the nonbreeding season (Jasnow et al. 2000; Scotti et al. 2008).

Studies in Syrian and Siberian hamsters gave rise to the third hypothesis. It was suggested that in these species, aggression is controlled by melatonin per se (Demas et al. 2004; Jasnow et al. 2002; Wang et al. 2012a, b). It remains to be elucidated how melatonin affects aggression. One series of evidence suggested that melatonin downregulates neuronal nitric oxide synthase activity independent of gonadal steroid hormones; according to this hypothesis, melatonin increases aggression in the nonbreeding season by reducing brain nitric oxide levels (Bedrosian et al. 2012; Wen et al. 2004). It is possible that the effects of melatonin are supported by testosterone synthesized from dehydroepiandrosterone under the effects of melatonin (Demas et al. 2004; Scotti et al. 2008, 2009). Taken together, aggressive behavior is independent of testosterone in certain species of hamsters, where aggression seems to be controlled by melatonin, likely by its effects on both nitric oxide synthesis and the conversion of adrenal steroids into testosterone. The consequences of this line of research for understanding aggression—beyond certain hamster species—are dual. Firstly, it directed attention towards the role of nitric oxide in aggression (Nelson et al. 2006); secondly, and more importantly for the present work, it revealed that the control of aggression may be overtaken by melatonin under certain conditions. Although this phenomenon received little attention in human research, some reports suggest that the aggression-promoting effects of melatonin are present in humans. For example, demented patients treated with melatonin showed increased aggression (Haffmans et al. 2001). Melatonin secretion may also be involved in self-injurious behavior (Herman 1990).

Taking into account the above, the response to the question asked in the title of this section is dual. Yes, testosterone affects aggression by interacting with brain mechanisms that control aggression. No, testosterone is not the only mechanism that controls aggression; moreover, there are contexts that render aggression independent of testosterone production. Under the influence of these contexts, alternative mechanisms are activated that enable aggressiveness to be expressed at high levels despite low levels of plasma testosterone.

2.2.3.3 Effects in Models of Abnormal Aggression

Testosterone seems to play a role in mice selected for aggressiveness. Short attack latency mice show higher levels of testosterone than long attack latency mice (the timid selection line) (van Oortmerssen et al. 1992; Compaan et al. 1993a, b, 1994a, b). Similar findings were obtained in Turku aggressive mice and cows selected for fighting ability (Plusquellec and Bouissou 2001; Sandnabba et al. 1994). The impact of testosterone seems weaker than that of genetic selection, because castration did not cancel genetically determined behavioral differences; however, aggression was decreased more strongly in short attack latency mice, suggesting that testosterone has a role in their abnormal aggression (van Oortmerssen et al. 1987). Subsequent research showed that the perinatal organizational effects of testosterone and adult brain aromatase activity are also altered in selection lines, further supporting the notion that abnormal aggression in these lines has a testosterone component (Compaan et al. 1992, 1993a, b, 1994a, b; Sandnabba et al. 1994; Sluyter et al. 1996).

The interaction between testosterone and abnormal aggression was also studied in the alcohol model of abnormal aggression. Two findings deserve attention: (1) the early organizational effects of testosterone and the effects of alcohol on aggression seem to interact because neonatal testosterone administration increased, while neonatal castration decreased the effects of alcohol on aggression in adulthood (Lisciotto et al. 1990; Winslow et al. 1988); (2) in adult animals (not treated in the neonatal period), there was a significant positive interaction between plasma testosterone and the effects of alcohol on aggression albeit this interaction was somewhat less robust (DeBold and Miczek 1985; Winslow and Miczek 1988).

In the case of other models, the role of testosterone in abnormal manifestations of aggression was not directly investigated (e.g., by castration or testosterone administration). Noteworthy, however, testosterone levels were measured in some of the models. Testosterone production was not altered in the maternal separation, postweaning social isolation, rat and hamster early subjugation, low-anxiety, and anabolic-androgenic models (Akbari et al. 2008; Amstislavskaya et al. 2013; Cunningham and McGinnis 2008; Ferris 2003; Ferris et al. 2005; Ortiz et al. 1984; Salas-Ramirez et al. 2008, 2010; Shimozuru et al. 2008; Veenema et al. 2006, 2007a, b; Veenema and Neumann 2009). In the high-anxiety model, basal levels remained unchanged, but the testosterone response to aggression was abolished (Veenema et al. 2007b). In the pubertal stress model, changes were seen not in testosterone levels per se but in the testosterone/corticosterone ratio increases (Márquez et al. 2013). In other models, e.g., the glucocorticoid dysfunction and repeated victory models, plasma testosterone levels likely increase but contradictory findings also exist (Balasubramanian et al. 1983; de Boer et al. 2003; Feek et al. 1989; Gao et al. 1996; Kalra and Kalra 1977; Poggioli et al. 1984).

In summary, the roles of testosterone were adequately studied in two models of abnormal aggression: in mice selected for aggressiveness and in the alcohol model. Testosterone appears to promote abnormal forms of aggression in both models. Indirect evidence suggests that the same is for the high-anxiety selection line, pubertal stress, glucocorticoid dysfunction, and repeated victory models, while a role for testosterone appears unlikely in other models. We mention, however, that this evidence is indirect.

2.2.4 *Does Testosterone Affect Aggression in Humans?*

2.2.4.1 Cons

The role of testosterone in human aggression is questioned and contested based on a series of arguments as follows: (1): females readily show aggressiveness despite their low testosterone levels; (2) aggression is not increased at puberty when testosterone levels increase; (3) high- and low-aggression individuals do not consistently differ in serum testosterone; (4) aggression does not increase in hypogonadal males when exogenous testosterone is administered to support sexual activity; (5) castration or antiandrogen administration to males is not associated with a consistent decrease in aggression; and (6) the replication of positive findings is often difficult. This set of contra arguments was formulated based on Albert et al. (1993); the reviews by Archer (1991) and Book et al. (2001) were also used. While studies that attest the role of testosterone in human aggression cannot be ignored, and these show that testosterone does control aggression, all the contra arguments listed above are valid.

Although less aggressive than males overall, females readily engage in aggressive conflicts in a variety of situations (*aggressive response to hypothetical conflict situations*: Reinisch and Sanders 1986; *aggressive responses in the Taylor Aggression Paradigm*: Böhnke et al. 2010a, b; *bullying in prisons*: Archer et al. 2007; *criminal violence and homicide*: Heide and Solomon 2009; *dating violence*: Cercone et al. 2005; *intimate partner violence*: Williams et al. 2008; *substance dependence-related aggression*: Bacskai et al. 2011). Gender differences were significant in most studies, but were not large. For example, the physical aggression score on the Buss-Perry Aggression Scale was 16.92 ± 0.69 in males and 14.63 ± 0.63 in females; gender differences were even smaller in substance-dependent subjects (males, 23.43 ± 0.49; females, 22.77 ± 0.89). In the study by Archer et al. (2007), males scored significantly higher on the physical aggression score of the Response to Victimization Scale (10.51), but the score of females was also considerable (7.58). Moreover, the score for displaced physical aggression was somewhat (nonsignificantly) higher in females (7.53) than in males (7.09). In the study by Böhnke et al. (2010a, b), unprovoked males and females showed comparable levels of aggression in the Taylor Aggression Paradigm; when provoked, males responded more aggressively, but females also increased their aggressive responses. Moreover, dating-related physical assault was significantly more frequent in female than in male undergraduate students (Cercone et al. 2005). Thus, females are almost as aggressive as males; this assumption is supported by meta-analyses showing that gender differences in aggression are consistent but are moderate in magnitude; the small effect size holds true for both verbal and physical aggressions (Hyde 2014).

The pubertal raise in testosterone production is accompanied by a rise in aggression in animals but not in humans. Moreover, physical aggression decreases in puberty as compared to childhood. While children primarily use physical aggression to resolve disputes, they gradually shift their behavior towards verbal (intimidation) and relational aggression to avoid the risk of retribution (Bjoerkqvist et al. 1992; Tremblay 1999). As a consequence, the positive correlation between testosterone and aggression seen in 12–13-year-old boys is lost by the age of 15–16 (Turner 1994), and plasma testosterone is linked to social dominance but not to physical aggression in early adolescence (Schaal et al. 1996).

Regarding the correlation between testosterone levels and aggression in males, we will avoid a "battle of references" (the counting of positive and negative findings) by taking examples from studies where the correlation *was* found significant. A common characteristic of such studies is that while the correlation is significant overall, individual correlation points are rather dispersed; consequently, some subjects are aggressive on the background of low testosterone, while other subjects show low aggression scores despite elevated plasma levels of testosterone. Such dispersed correlations or highly overlapping plasma testosterone ranges were found in a number of paradigms (*aggressive delinquency*: Banks and Dabbs 1996; *Buss-Durkee Hostility Inventory*: Ehrenkranz et al. 1974; *Point Subtraction Aggression Paradigm*: Carré et al. 2010; *reactive aggression*: Benderlioglu et al. 2004; *response to anger faces*: Wirth and Schultheiss 2007; *venturesomeness in personality disorders*: Coccaro et al. 2007a, b).[1] The conclusion deriving from this analysis is supported by the meta-analysis of Book et al. (2001): testosterone is a significant but weak contributor of aggressive behavior. Finally, a number of studies attest that increased testosterone production in women (due, e.g., to polycystic ovarian syndrome), and hypogonadism in males, and testosterone-based corrective treatments have no significant effects on aggressiveness (Finkelstein et al. 1997; O'Connor et al. 2002; Shufelt and Braunstein 2009; Stanworth and Jones 2008; Weiner et al. 2004).

These are serious arguments in favor of the idea that the role of testosterone in human aggression is minimal. Nevertheless, a substantial body of evidence demonstrates that under particular conditions, the effects of testosterone are rather strong (see Sect. 2.2.5). This conflict prompted a series of speculations on how this endocrine factor may affect aggression strongly such that its overall effects remain small.

2.2.4.2 Theories

A simplistic approach would suggest that the conflict is explained by errors in investigation. However, the large number of studies and the high scientific level at which many of these were performed excludes this possibility. The volatility of findings is not due to the volatility of methodologies, even if there was a permanent need to improve these. It occurs that the reason of contradictions lies in the complexity of the phenomenon. In fact, the question is not whether testosterone controls aggression or not; the real question asks about the conditions and circumstances that make testosterone operative or ineffective.

The Defensive Aggression Hypothesis The first animal studies substantiating the hypothesis were performed by Albert et al. (1986), while the theory was formulated by Albert et al. (1993). Animal observations suggested that in contrast to rivalry aggression which is typical to males, defense against predators is shown by both genders and does not depend on testosterone. In its final form, the theory postulated that human aggression is not influenced by testosterone because it resembles defensive aggression in animals (Albert et al. 1993). This resemblance is not very strong, however; defense against predators may model fear responses better than interpersonal conflicts (Blanchard and Blanchard 1989). Nevertheless, one can still assume that in some cases, human conflicts elicit fears similar to those felt by rodents faced with predators, and the behavioral response depends neither on gender nor on testosterone. This assumption explains why aggression *does not*

[1] Unfortunately, investigators often provide statistical figures (e.g., correlations coefficients) without illustrating the data. We cited here papers where correlations were shown on graphs. We note, however, that the phenomenon is general as judged from correlation coefficients provided in other papers.

depend on testosterone in some cases. Additional theories are needed to explain cases where it *does*.

The Challenge Hypothesis This hypothesis was put forward by Archer (2006) and is in fact a human extension of the hypothesis proposed by Wingfield et al. (1990) for monogamous birds. It was also influenced by a study in primates (Muller and Wrangham 2004). The hypothesis suggests that testosterone is increased by and promotes aggression under conditions of social challenge in a manner very similar to that seen in birds. It was also suggested that testosterone influences the trade-offs between momentary interests (sexual stimulation, competition, reputation, etc.) and parental investment: it promotes an extraverted, uninhibited, dominant personality that pursues shorter-term reproductive strategy, emphasizing mating rather than parental effort (Archer 2006).

The Developmental Hypothesis The observation that neonatal testosterone injections increase adult aggressiveness in mice (Bronson and Desjardins 1968, 1969) was later followed by human studies demonstrating that the same happens in people (Auyeung et al. 2009; Bailey and Hurd 2005; Benderlioglu and Nelson 2004; Hines et al. 2002; Reinisch 1981; Yildirim and Derksen 2012a, b). These studies showed that all three prenatal androgenic treatments, high amniotic fluid testosterone, and low 2D:4D digit ratios (indicative of high fetal testosterone) increase aggressiveness later in life. Albeit not formally advanced as a hypothesis so far, these findings (replicated by an increasing number of studies) suggest that the long-term neural effects of early testosterone exposure make adult aggressiveness somewhat independent of testosterone secreted in adulthood. In other words, normal adult levels of this hormone promote aggression by virtue of sensitization produced during the fetal/neonatal period. Consequently, not adult, but perinatal levels of testosterone set the propensity to behave aggressively, for which the measurement of adult plasma levels may be misleading regarding the relationship between testosterone and aggression.

"Cofactor" Hypotheses A series of findings suggest that the behavioral relevance of individual differences in testosterone production is increased or made noticeable by the co-occurrence of other conditions. Some of the authors adhering to this approach emphasize dual influences (*testosterone-MAO_A polymorphisms*: Sjoberg et al. 2008; *testosterone-plasma cortisol*: Dabbs et al. 1991; Terburg et al. 2009; *testosterone-noradrenaline, testosterone-serotonin, or testosterone-dopamine neurotransmission*: Chichinadze et al. 2010; Gerra et al. 2007; Pavlov et al. 2012; Witte et al. 2009). The common denominator of such findings (some being formally incorporated into hypotheses; Terburg et al. 2009) is the assumption that while testosterone per se may have a minor impact, it makes a difference if associated with low cortisol production, low serotonin neurotransmission, etc. This hypothesis also explains contradictory findings and suggests that the role of testosterone may be understood in the context of another deficiency. Other authors assume that the role of testosterone can be understood in the context of multiple imbalances. Hypotheses of this kind suggest that testosterone is only one piece of a mosaic of

features that lead to aggression. The most elaborated form of this hypothesis is the triple imbalance hypothesis of reactive aggression by van Honk et al. (2010), suggesting that reactive aggression is predicted by an imbalance between testosterone and cortisol secretion, cortical–subcortical communication, and left-sided frontal brain asymmetry that results from low serotonergic function. Although not elaborated to a comparable degree, the embedding of testosterone effects into a framework of multiple neural/endocrine deficits was suggested by a series of other authors (for examples, see, e.g., George et al. 2001; Mong and Pfaff 2003; Nelson 2005; Pavlov et al. 2012). These hypotheses suggest that testosterone is one of the multitudes of factors that contribute to the expression of aggression.

"Social Context Hypothesis" Some authors assume that the effects of testosterone are strongly modulated by social factors (Booth and Osgood 1993; Dabbs and Morris 1990; Fang et al. 2009; Kreuz and Rose 1972; Sluyter et al. 2000; Soler et al. 2000). In these studies, socioeconomic status, cultural background, and family relationships affected the testosterone response to conflicts and also the interaction between testosterone and aggression. Noteworthy, social context affects the impact of testosterone also in animals (Gleason et al. 2009; van Anders et al. 2011), suggesting that the interaction between testosterone, aggression, and context is a general phenomenon.

Taken together, these theories suggest that the testosterone-aggression relationship cannot be expected to be linear. Studies where no association between testosterone and aggression was found may fall short at considering conditions that influence testosterone efficacy. In other words, negative findings may have turned out positive if conditions other than testosterone were also taken into account.

2.2.5 *Testosterone and Abnormal Aggression in Humans*

High testosterone levels were associated with both criteria based on which normal and abnormal forms of aggression are delimited in humans: rule-breaking violence and aggression-related psychopathologies. In the first large-scale study, a sample of 4,462 US military veterans showed a significant correlation between testosterone and a variety of antisocial behaviors (Dabbs and Morris 1990). In a series of subsequent studies, plasma testosterone levels differentiated normal controls from violent offenders as well as nonviolent offenders from violent offenders (Aluja and García 2007; Banks and Dabbs 1996; Brooks and Reddon 1996; Booth and Osgood 1993; Chichinadze et al. 2010; Dabbs and Morris 1990; Dabbs et al. 1991; Fang et al. 2009; Kreuz and Rose 1972; Popma et al. 2007a, b; Rasanen et al. 1999; Terburg et al. 2009; Virkkunen et al. 1994a, b). The relationship was moderated by a variety of conditions in some of these studies (*cortisol*: Dabbs et al. 1991; Terburg et al. 2009; Popma et al. 2007a, b; *noradrenaline*: Chichinadze et al. 2010; *psychopathology*: Rasanen et al. 1999; *serotonin*: Virkkunen et al. 1994a, b; *sex hormone-binding globulin*: Aluja and García 2007; *social factors*: Booth and

Osgood 1993; Dabbs and Morris 1990; Kreuz and Rose 1972; Fang et al. 2009). Nevertheless, a role for testosterone in violent delinquency was clear, even if this role was not played alone or was not independent of other factors.

High testosterone levels were also found in adolescent boys showing externalizing problems (Maras et al. 2003), conduct or oppositional-defiant-disordered boys and girls (Pajer et al. 2006; Kirillova et al. 2008), antisocial personality-disordered adults (George et al. 2001; Lindberg et al. 2003; Rasanen et al. 1999; Sjoberg et al. 2008; Virkkunen et al. 1994a, b), borderline personality-disordered women (Roepke et al. 2010), as well as psychopathic patients (Glenn et al. 2011; Stalenheim et al. 1998; Terburg et al. 2009; Yildirim and Derksen 2012a, b). Again, these effects were moderated by a series of other conditions (alcohol dependence, MAO_A polymorphisms, serotonin functioning, severe conduct disorder in adolescence, sex hormone-binding globulin in plasma, and social factors), but often showed correlations with symptom severity. In a few studies, increased adrenal androgen (androstenedione and/or dehydroepiandrosterone) rather than testosterone production was associated with conduct and oppositional-defiant disorder in children (Dorn et al. 2009; van Goozen et al. 1998a, b, 2000a, b). Taking into consideration, however, that these adrenal steroids are converted into testosterone and estrogens in the brain, these findings suggest alternative routes of brain androgen supply at early ages when the production of this hormone is low rather than negate the role of testosterone.

2.2.6 Overview

The question whether testosterone influences aggression can be broken down into several sub-questions, the answers to which are all affirmative: perinatal testosterone positively affects the development of brain mechanisms that control aggression; testosterone in adulthood promotes aggression; the mechanisms of abnormal aggression have a testosterone component; and all these mechanisms are operational in both animals and humans. The major question unanswered by these statements is why the association between testosterone and aggression is so weak when investigated in large human samples?

We propose the following explanations:

1. The mechanisms mediating testosterone effects are complex, and changes in one variable may provide insufficient information on their functioning. For example, normal levels of testosterone production can easily promote aggression if the brain mechanisms mediating its effects are "over-primed" perinatally.
2. While brain mechanisms involved in the control of aggression are upregulated by testosterone, decisions regarding behavior are not taken at hormonal levels. Testosterone does not activate brain mechanisms, but facilitates functions when particular brain mechanisms are activated by other factors, e.g., a social challenge. In this respect, we refer again to the theoretical paper by Archer (2006)

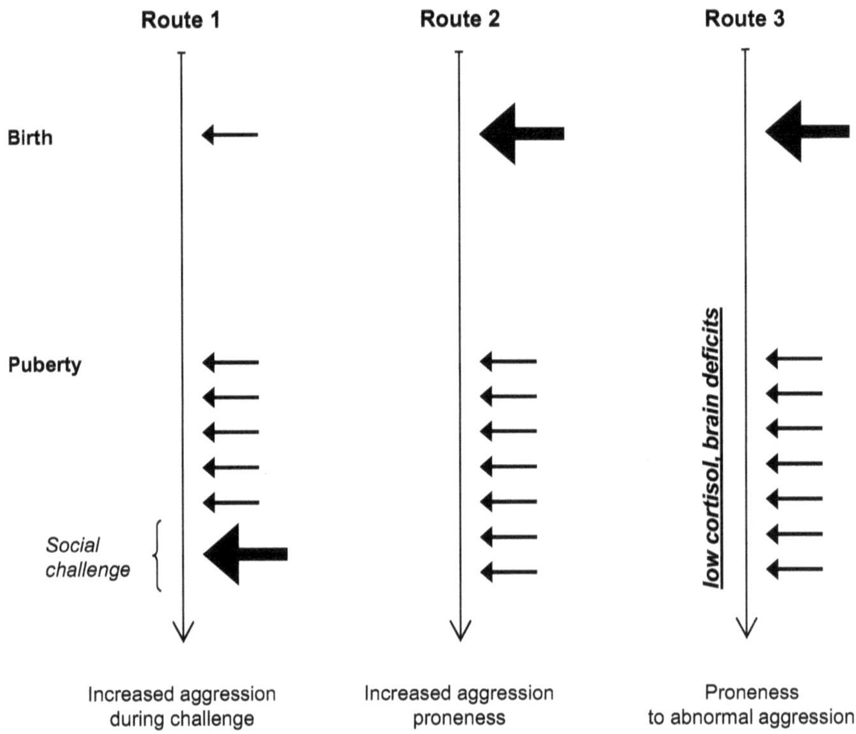

Fig. 2.4 *Three theoretical rutes by which various forms of aggression may develop despite normal basal testosterone levels.* The *arrows* indicate testosterone secretion; their length and thickness depicts the amount secreted. *Route 1*, challenge-induced increases in testosterone elevate aggressiveness temporarily (challenge hypothesis, Archer 2006); *Route 2*, increased prenatal exposure to testosterone elevates aggression-proneness overall; *Route 3*, a pattern similar to the previous one elicits abnormal forms of aggression when associated with other deficits (tripple inbalance hypothesis, van Honk et al. 2010). Note that testosterone influences aggression indirectly by affecting brain function

and also to Fig. 2.1. The interaction between challenges and testosterone functions may explain why effects are clearer in animals than in humans. In animals, the role of testosterone is studied in animals that are socially challenged (e.g., in the resident-intruder test). Correlational studies in humans may include people living in challenging environments but also include subjects who lead lives rather free of aggressive challenges.

3. The role of this hormone becomes especially conspicuous when associated with other mechanisms that promote aggressiveness (see, e.g., the triple imbalance hypothesis by van Honk et al. 2010). The activation of aggression-controlling brain circuitries by other factors inherently makes the roles of testosterone noticeable (see above). Interactions of this kind may explain why the role of testosterone is so strong in abnormally aggressive people who show multiple dysfunctions and so weak in healthy controls where testosterone is insufficient to activate aggression-bound brain mechanisms alone (Fig. 2.4).

Taken together, the above considerations suggest that the role of testosterone should be evaluated in conjunction with the complexity of its mechanisms of action and in conjunction with other factors that effect aggressiveness.

2.3 Glucocorticoids

Glucocorticoids control a wide array of physiological functions from blood glucose to immunity and brain mechanisms. The diversity of its actions forecasts some difficulties regarding the evaluation of the glucocorticoid/aggression relationship. This difficulty is increased by the complex secretion patterns of these hormones and by the multitude of its cellular and subcellular mechanisms of action. Its effects are further complicated by the fact that glucocorticoid effects are highly time dependent; effects occur and vanish on a time scale of minutes to month. The general characteristics of hormone action also need to be kept in mind: glucocorticoids do not activate behaviors but affect brain function. For these reasons, the analysis of the glucocorticoid/aggression relationship should be carried out carefully: the complexities of both aggressiveness and glucocorticoids should be taken into account.

2.3.1 Mechanisms of Action

In principle, two types of glucocorticoid action were identified: rapidly developing and transient effects mediated by a variety of cellular mechanisms and more slowly developing but persistent effects mediated by the genome. The non-genomic mechanisms endow glucocorticoids with the ability to rapidly affect brain function, by stimulating the voltage-dependent Ca^{2+} channel and calcium trafficking in general, by binding to and affecting the function of various neurotransmitter receptors, by binding to various membrane and cytoplasmic proteins, and by affecting protein–protein interactions within the neuron (for recent reviews, see de Kloet et al 2008a, b; Groeneweg et al. 2011; Haller et al. 2008). Genomic effects develop more slowly (within hours, days, weeks, and even month) and consist of the activation or inhibition of gene transcription. Such effects are rather wide ranging because glucocorticoid responsive elements are frequent in the genome, and in addition, glucocorticoids regulate gene expression also posttranscriptionally (Datson et al. 2008; Ing 2005).

Beyond large differences in lag time and cellular mechanisms, the two types of action also differ in the glucocorticoid secretion patterns that activate them, and in their ultimate outcome. Non-genomic effects are activated by acute surges in plasma glucocorticoids and are behavior activating in nature: they aim at controlling the critical situation that triggered the stress response (de Kloet et al. 2008a, b; Koolhaas et al. 2011; Haller et al. 1998a, b). This is valid for aggression as shown below. Non-genomic effects by contrast are typically activated by persistent changes in glucocorticoid secretion; overall, such effects are behavior depressing in nature and promote survival by limiting risks and costs associated with behavioral activity (Haller et al. 1998a, b). This is valid for aggressive behavior as shown below.

The particularities of glucocorticoid action require that its role in behavior control necessitates the detection of both acute surges and persistent changes. The first task is relatively easy because the timing of surges is made obvious by events, while their magnitude is large. Note that aggression is among the strongest stressors (Koolhaas et al. 1997). Uncovering persistent changes is rather difficult because glucocorticoid secretion shows diurnal, ultradian, and circahoral secretory rhythms, which are identifiable in both blood plasma and saliva samples (Follenius et al. 1987; Stupfel and Pavely 1990; Trifonova et al. 2013). In addition, they rapidly respond to meals and minor stressors and in addition the anticipation of stressors (Gaab et al. 2005; Tornatzky and Miczek 1994). Consequently, plasma levels are highly variable, and the timing of peaks and nadirs is unpredictable. This problem can be overcome by repeated sampling that washes out the effect of oscillations, by studying the so-called awakening response and by obtaining urine samples over prolonged periods (e.g., 24 h) (Fig. 2.5). The phrase awakening response covers the relatively rapid surge in glucocorticoid secretion that follows awakening. This is rather free of environmental influences and is considered the most appropriate measure of HPA-axis basal activity (Kudielka and Wüst 2010).

It is worth noting that persistent changes in secretion patterns are not restricted to the chronic stress response which involves a persistent increase in secretion. The possibility that the HPA-axis can chronically lower its activity together with the assumption that this condition leads to psychopathology was recognized by a study on violent antisocially disordered people (Virkkunen 1985) and received much attention in subsequent studies on abnormal forms of aggression. Compared to the conditions that chronically activate the HPA-axis, the factors that have the opposite effect are considerably less well known. This endocrine condition is generally attributed to early adversities (Caldji et al. 2000; Gunnar and Vazquez 2001; Saxbe et al. 2012; van Goozen et al. 2007). The same influences usually elicit chronic stress responses in most people, and only a minority of them develops the aforementioned reduction in HPA-axis function (Gunnar and Vazquez 2001). The distinctive features of this subgroup are unknown. Genetic predispositions, epigenetic phenomena triggered by gene–environment interactions, and the functional impairment of the adrenal cortex after birth complications or other dramatic stressors are suspected but unproven factors of developmental trajectories associated with chronic HPA-axis hypofunction (Rutter et al. 2006).

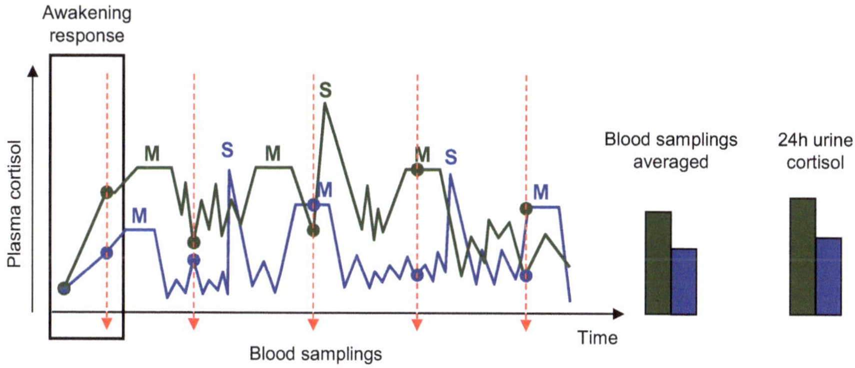

Fig. 2.5 *The interaction between glucocorticoid secretion rhythms and glucocorticoid levels measured at particular time-points. Left-hand panel,* an example on how the moment of blood sampling (*red arrows*) influence detected cortisol values (*filled circles*) in two individuals with clearly different cortisol secretion profiles (*blue and green*) depending on endogenous secretion rhythms, as well as minor stressors (*S*) and meals (*M*). *Right-hand panel*, Cortisol levels estimated by averaging multiple samplings and in urine collected over the whole day in the very same individuals. Note that variations during the night are poorly known, for which these were not represented on the graph

2.3.2 Normal Aggression in Animals

2.3.2.1 Acute Interactions

It was recognized very early that aggression is stressful and leads to glucocorticoid production in a variety of species (*fish*: Earley et al. 2006; lizards: Woodley et al. 2000; *mice*: Bronson and Eleftheriou 1965; *quails*: Ramenofsky 1985; *rats*: Schuurman 1980). In the rat resident-intruder test, the increase in glucocorticoids was similar in the two contestants at the beginning of the encounter, but values rapidly returned to basal levels in the winner while remaining high in the loser once dominance relationships were established.

Disparate early findings suggested that corticosterone secreted in response to the encounter might have an impact on ongoing behavior, as ACTH and corticosterone treatments increased aggressiveness while corticosterone synthesis inhibition decreased it (Brain et al. 1971; Bronson and Eleftheriou 1965; Munro and Pitcher 1985; Mainardi et al. 1987). Similar findings were obtained later in fish, where aggressiveness tightly correlated with corticosterone levels that preceded the aggressive encounter and acute treatment with glucocorticoid receptor antagonists decreased aggressiveness (Chang et al. 2012; Schjolden et al. 2009). In two parallel studies, Hayden-Hixon and Ferris (1991a, b) identified certain subregions of the hypothalamus as the main sites of the effect of corticosterone on attack behavior. In view of the fact that the hypothalamus is tightly involved in the control of biting attacks in all the species studied so far (Haller 2013), these findings suggest that corticosterone directly influences brain mechanisms that control attacks.

The existence of a "positive feedback loop" between aggression and HPA-axis activation was challenged in the early 1980s by the discovery of the genomic mechanisms of glucocorticoid action. The general assumption by that time was that glucocorticoids exert their effects only by the genomic mechanisms. As these mechanisms need time to take effect (at least 30 min but usually much more) and, in addition, the increase in glucocorticoids per se needs time (about 5 min), it was assumed that glucocorticoids cannot influence the aggressive acts that triggered their release, because by the time when plasma glucocorticoids increased and the genomic mechanisms were activated, most of the aggressive acts of dyadic encounters were already carried out. As a result, there remained nothing else to be influenced by glucocorticoids than submission, i.e., the consequence of losing the encounter. This belief was reflected in a series of studies, even in those where glucocorticoid synthesis blockade clearly decreased aggressiveness, and glucocorticoids increased aggressiveness (Munro and Pitcher 1985). Later on, a series of studies demonstrated that the rapid effects of glucocorticoids are real and several non-genomic mechanisms were also identified (de Kloet et al 2008a, b; Groeneweg et al. 2011; Haller et al 2008).

In a series of studies on the rapid interactions between glucocorticoids and aggression, we showed first that psychosocial encounters per se are sufficient to increase plasma glucocorticoids, i.e., the mere sensory contact with opponents is able to activate the HPA-axis (Haller et al. 1995). Similar findings were obtained in free-ranging animals as well (Landys et al. 2010; Silverin 1993). As aggressive encounters are preceded by a series of nonsocial and social behaviors that can be perceived as preparations for fights but are not aggressive themselves, these findings demonstrated that glucocorticoids are increased before the actual fights are started. In the following, we showed that natural variations in glucocorticoid synthesis, e.g., those related to ultradian and diurnal variations, are paralleled by changes in aggressiveness when rats are challenged in the resident-intruder test (Haller et al. 1998a, b, c, 2000a, b). Particularly, rats were significantly more aggressive in the increasing phase of their corticosterone fluctuation than their counterparts in the decreasing phase of their corticosterone fluctuations. Moreover, glucocorticoid enhancement-induced aggressiveness was normalized when the natural increase in glucocorticoid secretion was counteracted by glucocorticoid synthesis blockade or by glucocorticoid receptor antagonists. As the time frame of these hormonal and behavioral changes was less than 20 min, these findings indirectly supported the notion that glucocorticoids rapidly promote aggressiveness via non-genomic mechanisms. In the next step, we studied the interaction between glucocorticoid secretion and biting attacks that were elicited by the electrical stimulation of the hypothalamic attack area (Kruk et al. 2004). We found that the stimulation of the attack area dramatically increased corticosterone secretion while corticosterone decreased the electric threshold of attacks, i.e., facilitated the behavioral response. This suggested that there was a fast positive feedback loop between the adrenocortical stress response and brain mechanisms involved in aggressive behavior, the time frame of this interaction being approximately 10 min. Finally, we demonstrated that within 2 min, corticosterone administration restored aggressiveness that was downregulated by corticosterone synthesis inhibition (Mikics et al. 2004). This fast effect of corticosterone was resistant to protein synthesis inhibition, clearly demonstrating both the rapid aggression-promoting effect of glucocorticoids and the non-genomic nature of the mediating mechanism.

Taken together, the findings reviewed above show that (1) the HPA-axis is activated by encounters with unfamiliar individuals before the actual fights start and (2) glucocorticoids secreted in response to the encounter rapidly promote aggressiveness. Actually, aggressiveness is expressed concomitantly and under the supporting influence of glucocorticoids. In a subsequent study we showed that corticosterone does not affect aggressiveness in established colonies of rats, i.e., in colonies where the general level of aggressiveness is low and animals are not under the pressure of an acute social challenge (Mikics et al. 2007).

Thus, glucocorticoids rapidly and non-genomically promote aggressiveness specifically under conditions of a social challenge.

2.3.2.2 Chronic Effects

While acute surges in glucocorticoids promote aggressiveness in socially challenged animals, the effects of chronic increases in glucocorticoids are opposite.

As shown above, the aggressive encounter is stressful for both contestants; however, plasma glucocorticoids rapidly decrease after winning, but remain elevated in losers (*fish*: Höglund et al. 2002; *lizards*: Summers et al. 2003; *mice*: Bronson and Eleftheriou 1965; *monkeys*: Yodyingyuad et al. 1985; *rat colonies*: Blanchard et al. 1995; *rat dyadic encounters*: Schuurman 1980; *snakes*: Schuett and Grober 2000). In an experiment involving opponents with contrasting levels of aggressiveness, we showed that the effects of losing do not depend on the prefight status of individuals; plasma corticosterone levels were rapidly reduced in intruders that won the encounter, while the opposite happened in residents that were defeated in their home cage (Haller et al. 1996).

A series of studies show that elevated levels of glucocorticoids decrease aggression and promote submissiveness. This was demonstrated in a variety of species treated with glucocorticoids for prolonged periods (from 3 days to 3 weeks) (*decreased aggressiveness in fish, lizards, birds, and rats*: Meddle et al. 2002; Hayden-Hixson and Ferris 1991a, b; Overli et al. 2002; Tokarz 1987; *increased submissiveness, fish, mice, and rats*: DiBattista et al. 2005; Leshner et al. 1980; Hayden-Hixson and Ferris 1991a, b; *reduction of home-range lizards*: DeNardo and Sinervo 1994). Similar findings were obtained in experiments where subjects were exposed to stressors. Prolonged and strong single stressors (e.g., restraint for 2–6 h) or chronic social and nonsocial stressors decreased aggressiveness (Blanchard et al. 1995; Corum and Thurmond 1997; Wood et al. 2003; Yohe et al. 2012). The effects of post-encounter increases in glucocorticoids appear to be very strong and not necessarily related to the subjective experience of losing. For example, rats showed increased submissiveness if injected with corticosterone after an aggressive encounter on the previous day (Timmer and Sandi 2010). Thus, the primary mechanisms of losing are hormonal and not cognitive in rats.

Thus, durable increases in circulating glucocorticoids—which may or may not result from defeat experience—decrease aggressiveness and promote submissive behaviors.

2.3.3 Abnormal Aggression in Animals

Basal levels of glucocorticoids were not elevated in any of the abnormal models developed so far, which may not be surprising if one considers the consequences of chronic stress that were presented above. The glucocorticoid response to stress, however, was elevated in a number of studies, e.g., in the amphetamine, cocaine, instigation, early maternal separation, and postweaning social isolation models as well as in rats selected for low anxiety (Ansah et al. 1996; Fish et al. 1999; Moldow and Fischman 1987; Neumann et al. 2010; Toth et al. 2011; Veenema and Neumann 2009). The common feature of these models is that their subjects show exceedingly high levels of aggression. Other symptoms of abnormal aggression were investigated in one case only, namely, in the postweaning social isolation model, where increased glucocorticoid responses to aggression were associated with behavioral agitation, increased autonomic responses to aggression, and abnormal forms of attack (attack on vulnerable targets, deficits in social communication, and increased defensiveness on the background of increased attacks).

In another subgroup of models, either the basal levels of glucocorticoids were low or the glucocorticoid response to aggression was blunted (*subjugation model*: Ferris et al. 2005; *anabolic steroid model*: Rejeski et al. 1990; *rats selected for high anxiety*: Neumann et al. 2010; *mice selected for high aggressiveness*: Caramaschi et al. 2008a, b; Veenema et al. 2004). In the glucocorticoid dysfunction model, low basal glucocorticoid levels and blunted stress responses were experimentally induced (Haller et al. 2001, 2004). Where investigated, abnormal attack patterns were commonly seen; high attack counts were seen in some but not all models. No signs of behavioral agitation were observed, while autonomic responses to aggression or basal heart rates were diminished in those models where these were studied.

A third case is the pubertal stress model, where neither basal corticosterone nor the glucocorticoid stress response was altered, but the testosterone/corticosterone ratio increased, indicating a decreased corticosterone production relative to testosterone (Márquez et al. 2013).

The findings reviewed above demonstrate that abnormal aggression may be associated with either increased acute glucocorticoid responses to aggression or diminished HPA-axis functioning. The available data also suggest that subjects belonging to these two types of glucocorticoid/aggression associations also show differential emotional responses in terms of autonomic responses and may show some behavioral differences as well.

2.3.4 *Healthy Human Controls*

2.3.4.1 Acute Interactions

In the studies reviewed here, plasma cortisol was measured before and/or after the subjects performed a test where they executed aggressive acts. In this respect, these studies belong to that category of human studies which were called above "manifest aggression studies."

The delivery of punishment to imaginary opponents in a competitive situation increases plasma cortisol in subjects. For example, aggressiveness increased plasma cortisol in both the Taylor Aggression Paradigm—where subjects delivered punishment to opponents in the form of noises, the intensity, and duration of which were set by the subject (Böhnke et al. 2010a)—and the Point Subtraction Aggression Paradigm, where subjects can subtract points from opponents as a form of punishment (Gerra et al. 1997, 2001a, b, 2004, 2007; Geniole et al. 2011). These findings suggest that in these paradigms—the closest analogs of the resident-intruder test—the responses of animals and humans are similar and consist of a rapid increase in HPA-axis function. In contrast to animals, the impact of winning and losing was not studied in humans, as the chances of winning were standardized in these studies. Thus, this aspect of the glucocorticoid/aggression relationship is unclear in humans at present.

We identified two studies where aggressiveness was preceded by glucocorticoid administration or stress exposure. In the study by Böhnke et al. (2010b), subjects were treated with cortisol before the Taylor Aggression Paradigm. Cortisol increased aggressive responses but, interestingly, only in girls. In the study by Verona and Kilmer (2007), subjects were stressed by an air blast before they were allowed to deliver shocks to fictional employees. The pretest stressor increased the aggressive response in males, but not in females. These two studies may be perceived as contradictory by way of the gender differences obtained. Yet, one has to consider that the cortisol dose was low while the stressor was mild, and endogenous cortisol production was not suppressed. In addition, the subjects were inexperienced in the particular aggression-related task. In analogous animal studies (i.e., when endogenous glucocorticoid secretion was not suppressed and the subjects were fight inexperienced), glucocorticoids promoted locomotion without affecting aggression (Haller et al. 1997; Schjolden et al. 2009). Glucocorticoids increased aggression in such studies only when interference from endogenous secretion was eliminated. When compared to these animal studies, the aggression-promoting effects of cortisol and stress appear rather strong in the two human studies. We suggest that the contrasting gender differences were due to small sample sizes and to the lack of control over endogenous cortisol secretion.

There are several studies where the temporal relationship between the aggressive act and glucocorticoid stress responses was not as close as in the paradigms discussed above, but where the reciprocal facilitating role of aggression and glucocorticoid production seems likely. For example, amphetamine challenges increased plasma glucocorticoids more strongly in aggressive than in nonaggressive healthy subjects (White et al. 2006). Similarly, aggressive children showed increased cortisol responses to film viewing (Ivanov et al. 2011). These studies indicate that healthy subjects with a more reactive HPA-axis are more aggressive suggesting that aggressive children show increased HPA-axis responses in aggressive situations as well. In other studies, high cortisol levels at the time of measurement (usually after arrival to the laboratory) correlated with a higher number of aggressive acts performed in the following hours (Blair et al. 2005; Gunnar et al. 2010; Murray-Close et al. 2008). These correlations seem to be related to physical aggression as relational aggression showed the opposite correlation (Murray-Close et al. 2008). Also, the relationship may be valid for aggressiveness between peers, as acute changes in cortisol were not associated with aggressiveness in parent–child interactions (O'Neal et al. 2010).

Taken together, these findings show that the execution of aggressive acts increases glucocorticoid production in humans and this increase promotes their aggressiveness in a manner very similar to that seen in rodents.

2.3.4.2 Chronic Effects

In animals, defeat prolonged the glucocorticoid elevating effects of aggression and promoted submission. Similar findings were obtained in athletes where losers showed higher post-competition cortisol levels than winners, and a similar difference was seen on the long run between successful and less successful competitors (Booth et al. 1989; Parmigiani et al. 2009). By an interaction with testosterone, defeat-induced increases in cortisol result in decreased dominance and reluctance to compete again (Mehta and Josephs 2010). These findings are limited in number but suggest human/animal similarities in the effects of losing.

The effects of chronic glucocorticoid treatments were not studied in humans, likely for ethical reasons. However, cortisol elevates without experimental intervention in affective and mood disorders. Although anxious and depressed subjects cannot be considered healthy controls, they may still be considered in this section because their psychopathology is not directly related to aggression. Interestingly, a significant share (around 25 %) of depressed patients show outbursts of aggression called anger attacks (Fava 1998; Painuly et al. 2011). Social phobics are also prone to excessive violence under social pressure, including threatening with weapon, assault, and murder (Alden and Taylor 2004; Casiano et al. 2008). As both disorders are associated with chronically increased plasma cortisol, these findings suggest chronic cortisol—in marked conflict with animal findings—promotes aggressiveness. It remains to be seen whether this effect is secondary to their psychopathology (to neural changes induced by their state) or is due to the chronic elevation of cortisol production. Noteworthy, aggression seen in depressed and anxious patients occurs as a response to provocation, which may induce exacerbated HPA-axis responses in these patients. As such, aggressiveness in depression and social phobia may be a particular case of abnormal reactive aggression.

Taken together, these findings show that defeat increases plasma cortisol in athletes and decreases their willingness to engage in subsequent contests. In this respect, human and animal findings are rather similar. Aggression is increased under social pressure in a subgroup of depressed and socially phobic patients; the role of glucocorticoids in this phenomenon is unclear.

2.3.5 *Abnormal Aggression in Humans*

Studies in abnormally aggressive people focus primarily on the "glucocorticoid deficiency hypothesis of aggression," which proposes that abnormal aggression is associated with decreased cortisol production. Cortisol production is understood in terms of basal levels; in this respect, these studies are similar to those presented above under the heading "*chronic interactions*." The difference is that subjects are not healthy controls this time.

The first study on the subject was written by Virkkunen (1985). This paper showed that 24-h urinary free cortisol secretion was dramatically reduced in habitually violent offenders with antisocial personality disorder as compared to all three violent offenders without antisocial personality disorder, antisocial personality without the habitually violent tendency, and male clinic personnel. This seminal study prompted a series of investigations that partly confirmed, partly invalidated these early findings. Here we ask the question whether discrepancies are due to experimental error, to the volatility of the phenomenon, or to other factors relevant for aggression control.

The study by Virkkunen (1985) and most of the follow-up investigations belong to the "aggression history" type of investigations: subjects were categorized as aggressive or nonaggressive based on past actions; therefore, the chapter describing such reports will reiterate this phrase. A common characteristic of the studies included here is that they studied basal levels of plasma cortisol. The title of this section makes mention of this particularity. Acute cortisol responses were also studied in abnormally aggressive people; these will be presented in a separate section.

2.3.5.1 Aggression History Studies: Basal HPA-Axis Function

Studies in this section were grouped according the behavioral particularities of subjects. By and large, this grouping corresponds to the one presented in Chap. 1, except for externalizing behavior, which was not discussed there but still involves aggression-related behaviors to a certain extent. Within each group of studies, reports will be analyzed thoroughly regarding the technique of glucocorticoid measurements. This is justified by the complex nature of glucocorticoid secretion patterns as described in Sect. 2.3.1.

Behavior-Unspecific Studies: Externalizing

Externalizing behavior does include aggressiveness and also a series of other behaviors. Thus, this behavioral category is rather mixed as it regards its characteristics. A mechanical analysis suggests that findings are contradictory. While some studies did find a negative association between plasma glucocorticoids and externalizing behaviors (Cicchetti and Rogosch 2001; Ruttle et al. 2011; Shirtcliff and Essex 2008; Shirtcliff et al. 2005), others found no association (Klimes-Dougan et al. 2001) yet others found the inverse relationship: aggressiveness was associated with increased glucocorticoid secretion (Marsman et al. 2008).

Three of the studies detach from the others by the thoroughness of glucocorticoid measurements (Ruttle et al. 2011; Shirtcliff and Essex 2008; Shirtcliff et al. 2005). Glucocorticoids were measured repeatedly, and their averages were considered, which overcomes interference from ultradian oscillations. Moreover, "state"-like (momentary) and "trait"-like (temporally stable) differences in secretion patterns were differentiated in one of them (Shirtcliff et al. 2005). The overall conclusion of these studies is that concurrently measured externalizing problems and glucocorticoid measures do not necessarily correlate, but a correlation occurs on the long run. In the study by Ruttle et al. (2011), externalizing behavior predicted the development of flattened diurnal secretion rhythms, while in the other two studies, low glucocorticoid levels predicted increased symptom severity on the long run (Shirtcliff and Essex 2008; Shirtcliff et al. 2005).

Two studies are unsupportive of the negative association between glucocorticoids and externalizing (Klimes-Dougan et al. 2001; Marsman et al. 2008). As it regards glucocorticoid measurements, these studies are similar to that of Cicchetti and Rogosch (2001) who found that glucocorticoids were negatively associated with externalizing behaviors. All three studies investigated morning/awakening glucocorticoid levels; yet, their conclusions were sharply different. There was a difference, however, in the characteristics of the samples studied. While Cicchetti and Rogosch (2001) studied subjects with clinical-level externalizing problems, the severity of these problems was milder in the other two studies. Moreover, purely externalizing subjects were lacking from the sample of Klimes-Dougan et al. (2001) who compared primarily internalizing and mixed (internalizing/externalizing) children. The comparison of these findings may suggest that the association between glucocorticoids and externalizing behavior is restricted to subjects with severe behavioral problems.

Taken together, these findings suggest that the association between externalizing behavior and concurrent glucocorticoid secretion patterns is confusing. However, studies authored by an overlapping set of authors suggest that glucocorticoid levels and the severity of externalizing behaviors correlate on the long run: low glucocorticoids predict an unfavorable psychological developmental trajectory, while externalizing behaviors predict a glucocorticoid deficit on the long run. Tentatively, one can hypothesize that the contradictions in concurrent measurements are explained by the timing of measurements relative to the developmental trajectories. One can hypothesize that studies accidentally performed in early phases of the developmental trajectory did not, while studies performed at later phases did find a negative correlation between glucocorticoid production and externalizing behaviors. As such, apparently contradictory findings may in fact be due to timing differences and may be explained by a real interaction between "trait-like" glucocorticoid secretion patterns and behavioral developmental trajectories in terms of internalizing and externalizing behaviors.

Criminal Record Studies

In the study by Brewer-Smyth et al. (2004), highly violent subjects had low awakening cortisol; a similar interaction was found by Popma et al. (2007a) in subjects who committed petty offenses. Blunted diurnal cortisol rhythms were found in both. These data support the negative association between aggressive delinquency and lowered glucocorticoid production and extend the observations of Virkkunen (1985) to moderately aggressive delinquent populations and to awakening cortisol levels.

In studies where cortisol levels were measured over the day, low cortisol was not associated with aggressive delinquency but moderated the relationship between testosterone and aggressiveness. Particularly, low cortisol made the pro-aggressive effects of testosterone conspicuous (Dabbs et al. 1991; Popma et al. 2007b). In another study involving in male offenders with a history of violent acts, plasma cortisol was negatively associated with the affective-interpersonal features of psychopathy (Holi et al. 2006). The only study where findings were really conflicting from those reviewed above was published by Soderstrom et al. (2004). In this study, however, the circumstances of blood sampling suggest that stress responses rather than basal levels were measured; consequently, this study will be evaluated in the next section.

Psychopathology-Linked Studies: Mixed/Incomplete Diagnoses

The phrase mixed diagnosis stands here for "disruptive behavior" attributed to children that expressed a study-specific mixture of conduct, oppositional-defiant and attention deficit-hyperactivity symptoms, and "conduct problems," a mixture of conduct and oppositional-defiant symptoms. Although not always mentioned, subjects rarely met the conditions of full diagnosis for any of the diagnoses that made up their "disruptive" or "conduct" symptomatology.

Neither category of symptoms correlated with particular cortisol secretion profiles. Awakening and 24-h urinary cortisol were similar in disruptive and control children (Fairchild et al. 2008; Kruesi et al. 1989). When cortisol was measured in later phases of the day, some studies showed higher, others lower, yet other similar levels of basal cortisol in disruptive and control children (Fairchild et al. 2008; Kariyawasam et al. 2002; Scerbo and Kolko 1994).

More consistent findings were obtained when measures of aggressiveness were considered in addition to overall diagnosis; the negative correlation was significant in aggressive subgroups of disruptive children (Poustka et al. 2010; Oosterlaan et al. 2005; Van de Wiel et al. 2004). Callous-unemotional traits had a similar effect (Loney et al. 2006). Long-term associations were also revealed. Reliably low basal cortisol levels (averages of multiple measurements) predicted overt aggression and disruptive symptomatology on the long run (McBurnett et al. 2000; Sondeijker et al. 2008).

Psychopathology-Linked Studies: "Pure" Diagnoses

Both conduct and oppositional-defiant disorders were associated with low midmorning cortisol levels (Pajer et al. 2001, 2006; van Goozen et al. 1998a, b; Vanyukov et al. 1993). The same holds true for subpopulations where aggressive symptoms of the same disorders were particularly strong (Golubchik et al. 2009; Pajer et al. 2006; van Goozen et al. 1998a, b; Vanyukov et al. 1993), with the following specifications: the dehydroepiandrosterone/cortisol ratio rather than cortisol levels per se showed the expected correlation in the study by Pajer et al. (2006); in the study by Vanyukov et al. (1993), the association involved symptom counts without reference to the nature of the symptoms counted or their severity. Psychopathic symptoms were also associated with low awakening or midmorning corticosterone (Cima et al. 2008; Dolan et al. 2001; Vaillancourt and Sunderani 2011). Low cortisol was also seen in aggressive alcoholics (Bergman and Brismar 1994).

Azar et al. (2004) found no correlation between midmorning cortisol and conduct disorder-associated aggressiveness; however, the subjects of this study were pregnant girls. As pregnancy brings about large changes in cortisol production, this finding seems less relevant for the issues discussed here. The findings by van Bokhoven et al. (2005) are also at variance with the findings presented above; however, these authors studied stress responses rather than basal levels; consequently, this study will be evaluated in the next section.

Behavior-Centered Studies

There are a few papers where basal cortisol levels were associated with aggressive behavior but no diagnoses were mentioned. In theory, these studies should have been discussed in the "healthy controls" section. However, the populations investigated included individuals who showed very high levels of aggressiveness and, in addition, underwent highly stressful experiences (lived in refugee camps or suffered early maltreatment). As such, these subjects were not likely to be healthy in psychological terms and can be considered "abnormal" as it regards their aggressiveness.

Low cortisol was associated with higher aggressiveness in three studies involving highly aggressive albeit non-referred subjects (Gordis et al. 2006; Victoroff et al. 2011; Yu and Shi 2009). In one study, developmental aspects were also investigated (Shoal et al. 2003). In this interesting study, low cortisol levels (averaged point measures) at the age of 10–12 years predicted aggressiveness measured 5 years later.

No association was found in the study by Van den Bergh et al. (2008). According to the authors, however, no norms were available for the inventory they used. Therefore, it is unclear whether this study included subjects with unusually high levels of aggressiveness.

Evaluation

The studies reviewed here lead to a series of interesting conclusions: (1) while findings are incongruent as far as diagnoses are concerned (e.g., "externalizing," "disruptive"), a clear picture is provided by grouping subjects according to their aggressiveness. This suggests that cortisol correlates with aggressiveness and not with psychopathologies; (2) the diversity seen in animal models is not seen in humans, where abnormal aggression is invariably associated with low HPA-axis functioning; (3) low HPA-axis functioning predicts aggressiveness years ahead, which suggests developmental roles for the hormone. This issue was not studied in animals.

2.3.5.2 Acute Corticosterone Responses to Aggression

In a subgroup of these studies, plasma cortisol was measured either after the execution of aggressive acts or after exposing subjects to challenges that usually engender aggressiveness (e.g., provocation). These studies can be considered belonging to the "manifest aggression type." In their case, the glucocorticoid response seemed to depend on the type of aggression subjects were prone to. In situations likely evocative of reactive aggression, increased glucocorticoid stress responses were observed (*externalizing children*: Lopez-Duran et al. 2009; *hostile/reactive psychopathic subjects*: McBurnett et al. 2003, 2005; *children at risk for antisocial behavior and prone to reactive aggression*: O'Neal et al. 2010; *children showing antisocial traits*: van Bokhoven et al. 2005[2]). By contrast, subjects characterized by proactive forms of aggression showed blunted stress responses (*children with primary psychopathic traits*, McBurnett et al. 2003; *children at risk for*

[2] This author measured plasma cortisol immediately after the arrival of subjects into the laboratory. An overlapping set of authors (thus under likely similar conditions) showed that "arrival levels" are higher than those seen after a resting period (van Goozen et al. 1998a, b). As such, van Bokhoven et al. (2005) studied reactive rather than basal levels of plasma cortisol which explains their inclusion in this section.

antisocial behavior and prone to proactive aggression: O'Neal et al. 2010). Studies employing small sample sizes failed to show significant correlations of this kind, but the marginally significant correlations were in the expected direction (Van de Wiel et al. 2004). In a different set of studies, the challenge was not social but neurochemical. Subjects likely characterized by proactive forms of aggression (aggressive personality-disordered and psychopathic subjects) showed blunted cortisol responses to treatments regularly used to characterize HPA-axis responsiveness (Almeida et al. 2010; Dolan et al. 2001; Snoek et al. 2002).

There are a series of studies where the reactive-proactive distinction cannot be made or the stressor was unrelated to aggressiveness. Typically, the cortisol response of subjects was decreased in these studies (*aggressive delinquents*: Moss et al. 1995; Popma et al. 2006; *disruptive children*: Fairchild et al. 2008; van Goozen et al. 2000a, b; *attention deficit-hyperactivity disorder*: Yang et al. 2007; *heroin addicts*: Gerra et al. 2001b; Gerra et al. 2004). Increased stress responses were reported by one study on violent offenders (Soderstrom et al. 2004[3]).

Taken together, these studies show that the association between abnormal aggressiveness and acute stress responses is more varied than the association seen with basal levels. Note that in some cases, the stress factor involved the execution of either aggressive acts (e.g., point subtraction from partners) or feelings that engender aggression under nonexperimental conditions (provocation or frustration). Tentatively, reactive aggression was associated with increased, while proactive aggression with decreased stress responses.

2.3.6 Overview

Although the role of glucocorticoids in aggression is often perceived controversial, a detailed glucocorticoid and behavioral measure-sensitive analysis of the available findings suggests that this role is rather clear-cut; moreover, there is a marked similarity between animal and human findings. The conclusions of our analysis can be summarized as follows (Fig. 2.6):

1. Aggressive interactions rapidly increase glucocorticoid production, while glucocorticoids rapidly promote aggressiveness, suggesting the existence of a positive feedback loop between the execution of aggressive acts and acute glucocorticoid responses. This statement appears valid for normal aggression in both animals and humans.
2. Defeat results in chronically increased plasma glucocorticoids which suppresses animal aggression and promotes submission. The statement is valid for both

[3] The authors of this study sampled cerebrospinal fluid for cortisol measurements by lumbar puncture, which is a rather invasive procedure, and its anticipation likely increased cortisol production. In our view, these authors studied anticipatory stress, which explains the inclusion of their study in this section.

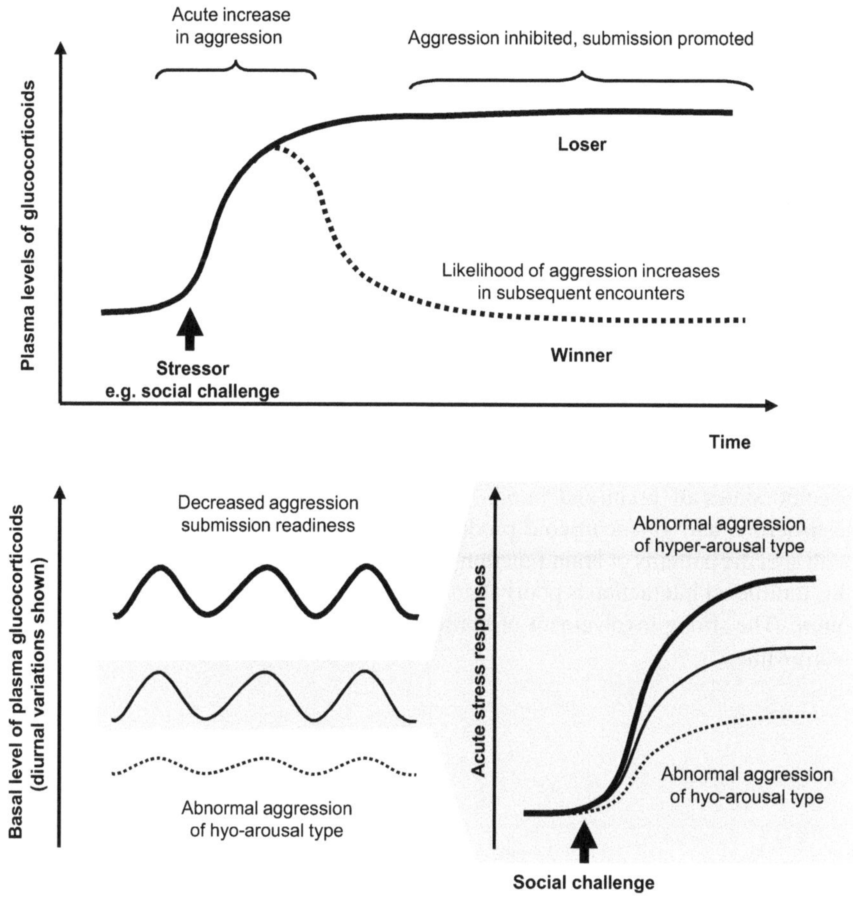

Fig. 2.6 *Intuitive representation of aggressive behaviors associated with particular glucocorticoid secretion profiles. Upper panel,* changes in plasma glucocorticoids and aggressiveness during an aggressive encounter in winners and losers. *Lower panels,* Developmental factor-induced changes in glucocorticoid secretion patterns and parallel shifts in aggressiveness. *Left*; changes in basal levels; *right*, acute response to aggression. Note that normal basal secretory rhythms may be associated with abnormal forms of aggression if acute stress responsiveness was altered

animals and humans, albeit the information is limited and somewhat controversial in the latter.

3. Abnormal aggression is associated with chronically low levels of glucocorticoid production in a subgroup of rodent models and in practically all types of abnormal human aggression.
4. Decreased stress responses were observed in a subgroup of abnormally aggressive people and in a subgroup of rodent models. Tentatively, these may be associated with proactive forms of aggression.

5. Increased stress responses were observed in a subgroup of abnormally aggressive people and in a subgroup of rodent models. Tentatively, these may be associated with reactive forms of aggression.

2.4 Overall Evaluation of Hormonal Determinants

The two hormones discussed here are clearly involved in the control of aggression, but the effects of neither can be described in simple terms. Nevertheless, their associations with aggressive behavior have a clearly recognizable pattern, which seems explanatory not only for aggression per se but for the forms that this behavior can take and the conditions under which these are expressed.

Hormone effects on behavior are not direct but are mediated by the neural changes induced by them. Formulated somewhat trivially, hormones determine specific states of brain and mind. While it is likely that particular patterns of testosterone and glucocorticoid production (and their interactions) are associated with specific patterns of brain function, which result in specific patterns of behavior, this multilevel interaction is poorly known, and its elucidation remains a task for the future. The strong involvement of hormones in behavior control makes such trials worthwhile.

Chapter 3
Neural Circuits Subserving Aggression: General Models

Any detail on the control of aggression is extremely valuable, but the grand aim is the understanding of the *circuitry* that subserves aggression and the identification of those alterations in the system that makes it dysfunctional. This major objective does not seem to be achieved at present stage. The main problem with the theories is their large number. While they operate with similar concepts overall, and cross-fertilization is also evident, approaches remain rather divergent. Our attempt of understanding aggression control will start with a brief presentation of the current concepts. Neither will be scrutinized at this stage. The following chapter will be devoted to an in-depth analysis of the nodal points of aggression control. The last chapter is a trial to put pieces together.

3.1 Animals

3.1.1 Cats

The most extensive mapping of the circuits that control aggressiveness was performed by the group of Siegel in cats; the studies available so far cover a period of 38 years (Siegel and Flynn 1968 through Bhatt and Siegel 2006). By using a complex set of advanced technologies, the group not only uncovered the brain areas that are involved in cat aggression and their connections but also the neurotransmitters used by the neurons of the circuit. A simplified summary of this exceptionally systematic work was summarized in Fig. 3.1.

The centerpiece of the circuitries identified by this group is the hypothalamus, the role of which was revealed by the seminal work of Hess (1928). In fact, the Siegel group asked two major questions: (1) which are the mechanisms that control the release of attacks from the hypothalamus, and (2) how does the hypothalamus control the sequence of movements that make up an attack? They identified two topographically distinct aggression-controlling mechanisms. One of these controls

J. Haller, *Neurobiological Bases of Abnormal Aggression and Violent Behaviour*,
DOI 10.1007/978-3-7091-1268-7_3, © Springer-Verlag Wien 2014

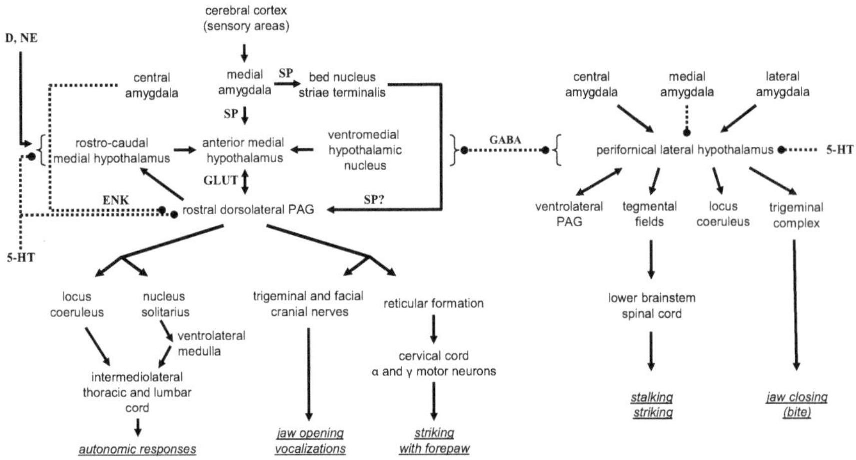

Fig. 3.1 *The circuitry that controls affective aggression (left) and predatory aggression (right) in the cat* (based on Siegel et al. 1999; Gregg and Siegel 2001; Siegel et al. 2007). *Abbreviations*: *5-HT* serotonin; *D* dopamine; *GABA* gamma aminobutyric acid; *GLUT* glutamate; *ENK* enkephaline; *NE* norepinephrine; *SP* substance P; *arrows, continuous line*, stimulatory contacts; *two-sided arrows*, bidirectional stimulatory contacts; *dotted lines with bullets*, inhibitory contacts; *dotted lines with bullets on both sides*, bidirectional inhibitory contacts

affective aggression and involves a downward stimulatory stream that includes the medial amygdala, medial hypothalamus, and dorsal periaqueductal gray, while the other controls predatory aggression and includes the central and lateral amygdala, lateral hypothalamus, and a series of downstream areas including the ventral periaqueductal gray. From upstream, both systems were assumed to be controlled by information received from the sensory cortex, while downstream, both were assumed to control neural mechanisms involved in autonomic and motor control. The two systems reciprocally inhibit each other at multiple levels, which implies that affective aggression inhibits predation, and similarly, affective aggression has low chances of expression during hunting. Stimulations within the circuitry were assumed to be realized by glutamatergic and substance P-ergic synapses, while inhibitions were assumed to be mediated by GABAergic and enkephalinergic neurotransmission (Fig. 3.1).

A blatant feature of the circuitry outlined by this group is the omission of the prefrontal cortex, which is considered a key region by practically all the other authors. The omission was not due to lack of interest or the incomplete survey of relevant subregions. Brutus et al. (1984) studied the involvement of this brain area by a combined technique. They electrically stimulated various prefrontal and other regions and identified the activated brain regions by [14C]2-deoxyglucose analysis.[1] In the first series of studies, stimulation was performed in the dorsomedial,

[1] This technology makes use of a radio-labeled, chemically transformed glucose molecule, which is readily uptaken by, but is neither metabolized within nor released from the neuron.

ventromedial, and infralimbic prefrontal cortices. Based on the figures provided, these areas seem analogous with the medial orbitofrontal, prelimbic and infralimbic cortices of rats. This detail appears important especially in comparison with rat findings. The stimulation activated a series of downstream areas, the largest activation being seen in the mediodorsal thalamus. No activation was observed in the hypothalamus. In the next step, the authors stimulated the mediodorsal thalamus, which activated the hypothalamus neither, but among other nuclei it activated the nucleus reuniens. When this was stimulated, finally the hypothalamus was activated; activation covered both the medial and lateral hypothalamus and a descending system, which could be followed to the level of the periaqueductal gray. This study suggested that information from the prefrontal cortex does reach the hypothalamus, but by a multi-synaptic pathway that goes through the mediodorsal thalamus and nucleus reuniens. The lack of the prefrontal cortex from all the schematics provided by this group is likely explained by the complexity of the synaptic links.

Another interesting feature of the mechanisms depicted in Fig. 3.1 is that affective aggression results in the opening of the jaws, while their closing results from the activation of the predatory mechanism. The likely reasons of this strange assumption are discussed below.

3.1.2 Rats

Hypothalamic mechanisms of aggression, including its neurochemical control, received considerable attention in rat studies as well (for a review see Kruk 1991), but the circuits controlling the hypothalamus and those translating hypothalamic activation into movement patterns were less systematically investigated. However, the circuits controlling rivalry aggression may be deduced from the findings of two studies that investigated the efferent and afferent connections of the mediobasal hypothalamus. Efferent connections were studied by Roeling et al. (1994), while afferent connections by Toth et al. (2010). Both studies found a surprisingly complex set of monosynaptic connections; the efferents were sent to and the afferents were received from over 50 brain areas, many of which overlapped. The large degree of overlapping demonstrates that most of the connections are reciprocal. Examples of these are shown in Fig. 3.2 (left-hand panel). There were a few notable exceptions from this reciprocity; it occurs that the connections are unilateral in the case of the prefrontal cortex-mediobasal hypothalamus and the mediobasal hypothalamus-periaqueductal gray connection; the medial amygdala-mediobasal hypothalamus connectivity appears to be highly asymmetrical. In all three cases, a downward stream of information can be

Consequently, activated neurons (those which were made to "work" by the stimulation) can be identified by their radioactivity. Note that energy production relies exclusively on the uptake from blood and degradation of glucose in neurons.

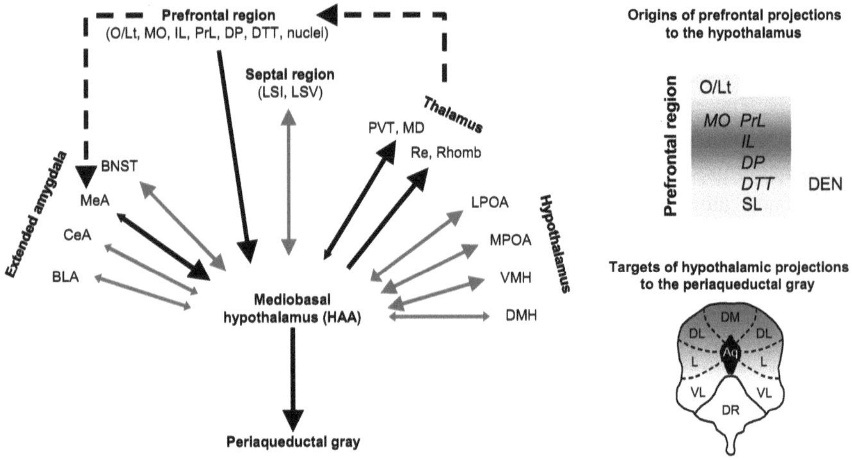

Fig. 3.2 *Afferent and efferent connections of the mediobasal hypothalamus in the rat* (based on Roeling et al. 1994 and Toth et al. 2010). *Legend*: *Direction and size of arrowheads*, represents the direction and stregth of the connection; *Black arrows*, connections that suggest directionality; *gray arrows*, reciprocal connections; *dashed arrows*, connections evidenced by other studies; *grayness in right-hand panels*: the strength of connections. Explanations for abbreviations were grouped according to major brain areas. *Prefrontal regions: DEN* dorsal endopiriform cortex; *DP* dorsal peduncular cortex; *DTT* dorsal tenia tecta; *MO* medial orbitofrontal cortex; O/Lt olfactory-limbic transition area; *PrL* prelimbic cortex; SL semilunar nucleus; *Septal region*: *LSI* lateral septal nucleus, intermediate part; *LSV* lateral septal nucleus, ventral part; *Extended amygdala*: *BLA* basolateral amygdala; *BNST* bed nucleus striae terminalis; *CeA* central amygdala; ; *MeA* medial amygdala; *Thalamus*: *MD* mediodorsal thalamic nucleus; *PVT* paraventricular nucleus of the thalamus; *Re* nucleus reuniens; *Rho* rhomboid nucleus; *Hypothalamus*: *DMH* dorsomedial hypothalamic nucleus; *HAA* hypothalamic attack area; *LPOA* lateral preoptic area; *MPOA* medial preoptic area; *VMH* ventromedial hypothalamic nucleus; *Periaqueductal gray*: *DM* dorsomedial nucleus; *DL* dorsolateral nucleus; *L* lateral nucleus; *VL* ventrolateral nucleus; *DR* dorsal raphe

assumed. This downward information flow may be complemented by a prefrontal cortex-medial amygdala connection, which was not studied by the aforementioned authors, but is evident from other sources (e.g., McDonald et al. (1999) described a direct connection between particular infralimbic neurons and the medial amygdala).

Taken together, these findings suggest that the affective aggression mechanism evidenced in cats does work in rats and is composed of the same neural structures: the prefrontal cortex, medial amygdala, mediobasal hypothalamus, and periaqueductal gray. Other similarities between the two species are also suggestive of shared mechanisms. For example, the medial hypothalamus appears to be controlled by the same prefrontal subareas (the orbitofrontal, prelimbic, and infralimbic cortices) in cats and rats, even if the control seems indirect in the former and direct in the latter. Moreover, affective aggression is controlled by dorsal aspects of the periaqueductal gray in cats, while the hypothalamic attack area of rats preferentially projects to this region in rats (see the right-hand panels of Fig. 3.2).

In addition to this downward stream, asymmetric connections suggest the existence of an upward flow of information. The components of this putative stream are the mediobasal hypothalamus, various nuclei of the thalamus, and the prefrontal cortex (for evidence on thalamus-prefrontal cortex connections, see Groenewegen 1988). The assumption that the mediobasal hypothalamus sends information to the prefrontal cortex via the thalamus is supported by a study where the electrical stimulation of the hypothalamic attack area was followed by the activation of the prefrontal cortex despite the lack of a direct anatomical link; this was explained by the mediatory role of the mediobasal thalamus, which was also activated by stimulation (Halasz et al. 2002b).

Based on the above and sporadic data on predatory aggression in rats, we hypothesized that the mechanisms controlling rodent aggression are very similar to those operative in cats: rivalry aggression is governed by a circuit composed of the medial amygdala, mediobasal hypothalamus, and dorsal periaqueductal gray, while predatory aggression is controlled by a parallel mechanism composed of the central amygdala, lateral hypothalamus, and ventral periaqueductal gray (Haller 2013).

3.2 Humans

Information on the neural control of aggression in animals (especially in cats) was used to build simple mechanistic models of human aggression, the various forms of which were perceived as variants of affective and predatory aggressions that are seen in cats (Siegel et al. 2007, Siegel and Douard 2011). According to these views, the mechanisms operating in animals and humans are very much the same with the exception that there is some room for free will in the latter. However, psychological approaches that can be traced back to the nineteenth and early twentieth century operated with quite different notions, and their impact is strongly felt even today. Current theories in fact amalgamate the two approaches; human aggression is usually understood by an interaction between psychological and neural factors. The general assumption is that the substrate of the former is the prefrontal cortex, while the substrates of the latter are brain areas similar to those that control aggression in animals.

The role of the prefrontal cortex in human aggression is usually explained in terms of its role in emotional and cognitive functions. Theories suggest that deficits in prefrontal functions engender aggressiveness by the loss of control over emotions and impulsivity, the inability to recognize emotions in others and lack of empathy, the inability to evaluate the consequences of actions (including insensitivity to punishment), and deficits in emotional learning which result in faulty decision making (Anderson and Kiehl 2012; Blair 2001, 2007, 2010a, b; Brower and Price 2001; Davidson et al. 2000; Gao et al. 2009; Golden et al. 1996; Herpertz and Sass 2000; Hoaken et al. 1998; Lopez et al. 2004; Raine 2002; Yang and Raine 2009). The role of hypoarousal and sensation seeking was also emphasized by some authors (Gao et al. 2009; Herpertz and Sass 2000; Lopez et al. 2004; Raine 2002). These psychiatric mediators of prefrontal deficits appear in different combinations and are emphasized in different ways by the authors referred to above. Yet their conclusions are consonant overall.

Other models of aggression address the role of the prefrontal cortex in more "biological" ways. The first trial of this kind suggested that aggression is controlled at three levels: the brainstem and hypothalamus, the limbic lobe (mostly the amygdala) that controls the hypothalamus, and the frontal neocortex, which modulates both limbic and hypothalamic output (Weiger and Bear 1988). Notably, this study explained the role of the prefrontal cortex in the context of cognitive factors; neural mechanisms were discussed in general terms only. Niedermeyer (1998) conceptualized the role of the prefrontal cortex in terms of the role played by this region in the initiation, planning, designing, and sequencing of movements that are performed in tight conjunction with working memory traces that are also hosted by the prefrontal cortex. This framework was used to speculate on the mechanisms of disinhibitory frontal lobe syndromes that range from tactlessness to aggression. The hypothesized information flow starts from the posterior cerebral regions (sensory information), reaches the prefrontal cortex through the mediodorsal thalamus, and assumes integration with locally stored working memory traces to give rise to cognitive processes associated with aggression. Herpertz and Sass (2000) attributed aggression-related behavioral dysfunctions to a faulty interaction between the ventromedial prefrontal cortex, the medial temporal cortex, and the amygdala, without entering into speculations regarding their networking. Lee and Coccaro (2001) emphasized the role of prefrontal serotonergic inputs. This framework attributed an indirect role to the prefrontal cortex; aggression was explained in terms of a serotonin deficiency-induced general behavioral disinhibition (e.g., impulsivity). The hypotheses by Blair (2001) differentiate the mechanisms underlying the two major types, namely reactive and instrumental aggressions. According to the views of this author, reactive aggression (behavior triggered by threat or frustration) is explained either by a faulty assessment of behavioral options and possible outcomes where somatic (e.g., autonomic) responses have an important role, or by deficits in social learning. The orbitofrontal cortex was assumed to play a role in both mechanisms. By contrast, instrumental aggression (behavior motivated by gain) was assumed to be driven mostly by the amygdala, which was considered a brain site of social conditioning. Lopez et al. (2004) also differentiated reactive and proactive aggression. They attributed reactive aggression to frontal asymmetries based on the assumption that the left and right prefrontal cortices are involved in the processing of positive and negative emotions, respectively. Reactive aggression was explained by an altered equilibrium between the two sides of the prefrontal cortex, while proactive (instrumental) aggression was attributed to a general decrease in prefrontal functions, which results in behavioral disinhibition. Schmahl and Bremner (2006) reevaluated the role of serotonin in borderline personality disorder-associated aggression and suggested that the effects of serotonin are mediated by the orbitofrontal and anterior cingulate cortices. Aggression was explained by diminished serotonin input to these areas—especially the anterior cingulate—that increases impulsivity and impulsive aggression.

Recently, several more elaborated neural models were proposed. These do make reference to emotions and cognitions, but basically they are mechanistic neural models. Berdahl (2010) assumes mostly based on cat data that "defensive aggression" (typically seen in borderline patients) is triggered by the dorsolateral column of the periaqueductal gray matter, a brainstem formation. Based on the way it is defined, the phrase "defensive aggression" is synonymous with the phrase "reactive aggression" mentioned above. The dorsolateral periaqueductal gray was assumed to be activated by the right-hand side basal amygdala. The effects of the basal amygdala are modulated by the lateral and central amygdalae, which were supposed to make a choice between two alternative responses, namely, dissociation (tonic immobility) that is mediated by the ventrolateral periaqueductal gray and aggression that is mediated by the dorsolateral periaqueductal gray. At a higher level, the functions of the right basal amygdala were supposed to be under the control of the right ventromedial prefrontal cortex and/or the right anterior cingulate. These two structures and the basal amygdala would reciprocally inhibit each other. As such, the expression of anger and aggression ultimately depends on the equilibrium of these reciprocal inhibitions. When the right basal amygdala is overactive and the right ventromedial prefrontal and/or cingulate cortices are hypoactive, the amygdala-periaqueductal gray pathway is released from inhibition, and the dorsolateral periaqueductal gray triggers aggression. A similarly detailed mechanism was envisaged by Blair (2010b) for "reactive" ("defensive") aggression. Although in many respects similar to the one proposed by Berdahl (2010), there are several major differences between the two models. Blair (2010b) attributes an important role to the hypothalamus, which was totally neglected by Berdahl (2010), despite the fact that the hypothesis of the latter was based on cat studies performed over decades by the group of Siegel, who considered the hypothalamus a centerpiece of aggression mechanisms. Blair (2010b) assumes that the basic mechanism of aggression is a network that includes the medial amygdala, the basal hypothalamus, and the dorsal half of the periaqueductal gray. Aggression occurs when stimulatory information runs through this pathway in the order in which the structures were listed above. An alternative (similarly aggression-promoting) route was proposed to involve the basolateral amygdala, central amygdala, and hypothalamus/PAG. This is another major difference between the two models, as Berdahl (2010) perceives the central amygdala as an aggression inhibiting center, while Blair (2010b) assumes the opposite. We stress that the two authors proposed models for the same type of aggression albeit in different contexts. Berdahl (2010) evaluates the mechanisms of reactive aggression in borderline personality disorder, while the model by Blair (2010b) focuses on reactive aggression shown by psychopaths. A third major difference between the two models is that Blair (2010b) detaches to a certain extent from a neural-mechanistic approach as far as the prefrontal cortex is concerned. The paper gives the impression that the control of aggression is perceived as being composed of two parts: a "mechanized" network consisting of the amygdala, hypothalamus, and periaqueductal gray and a cognitive/emotional component represented by the prefrontal cortex.

The major innovation put forward by Potegal (2012) is the integration of the temporal cortex in the mechanisms that control aggression. This author suggested that the prefrontal cortex receives environmental information not from subcortical sensory systems or the mediodorsal thalamus (as proposed by others) but from the posterior and middle temporal cortices. Moreover, he assumed that the incoming sensory information is integrated with memory traces at this level by an interaction between the aforementioned temporal areas and the anterior temporal cortex. As such, sensory information is integrated and processed in the temporal cortex before they reach the prefrontal cortex. The second major difference between Potegal's model and all the other models discussed here is the way in which the role of the prefrontal cortex is viewed. Based on rat findings published by Toth et al. (2010), Potegal (2012) suggests that there are direct connections between the ventromedial and orbitofrontal cortices and the hypothalamus; particularly, the former two inhibit hypothalamic mechanisms of aggression by stimulating local GABAergic neurons located within the hypothalamic area that are tightly involved in the elicitation of attacks. The existence of local GABAergic inhibitory neurons was also inferred from rat studies (Hrabovszky et al. 2005). According to Potegal (2012), the prefrontal cortex is the origin of two pathways: one inhibits aggression, while the other promotes it. The first one involves the indirect inhibition of the hypothalamic attack area by neurons located in the ventromedial and orbitofrontal cortices, while the second route involves the cingulate cortex and amygdala, which stimulate the hypothalamic centers of attack.

In conclusion, hypotheses are of three major types:

1. Theories of the first type assume that aggression is due to cognitive and emotional failures.
2. Type 2 theories converge on the general assumption that aggression emerges when "programmed" mechanisms located in the diencephalon and midbrain are released from the cognitive and emotional control of higher-order brain areas, especially the prefrontal cortex.
3. Theories belonging to type 3 assign neural circuits to both the psychological and behavioral aspects. In addition, they assume that "programmed" mechanisms residing in lower brain areas not only are released from inhibition but are actively controlled by higher-order brain areas, particularly by distinct mechanisms of the prefrontal cortex. The most recent and rather well-elaborated theory of Potegal (2012) was shown as an example of such theories in Fig. 3.3.

3.3 Outlook

It occurs that explanatory models of aggression have two main sources. Neurobehavioral models were inspired by the seminal work of the Siegel group, which on its turn was largely influenced by earlier work done by Hess, Flynn, and

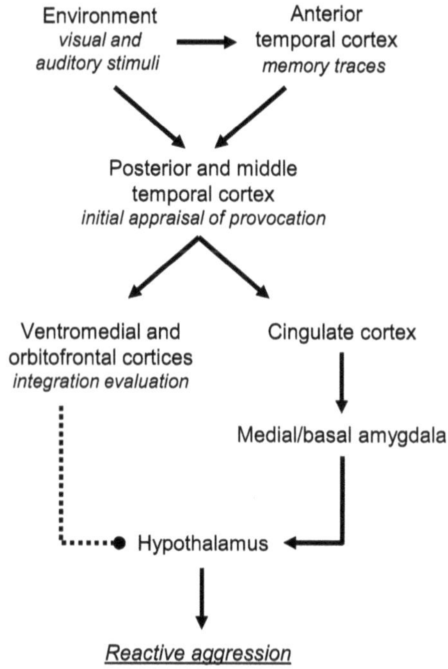

Fig. 3.3 *A representative recent example of type 3 theories on the control of human aggression* (based on Potegal 2012). *Abbreviations*: *normal font*, brain areas; *italic*, information processed; *underlined, italic*: behavior; *arrows, continuous line*, stimulatory contacts; *dotted lines with bullets*, inhibitory contacts; For other human models see Sect. 3.2

Bandler. This is valid for both rodent and human models of aggression. Siegel's findings were exploited in various ways, depending on the nature of other information that was available in these two species; nevertheless the basic template is recognizable in both. The other source of information was psychology, particularly, almost age-old views on the cognitive and emotional factors that engender aggressiveness. This reasoning was primarily employed in human models. Novel approaches complemented these views by assigning brain areas to psychological functions.

Albeit sharing basic concepts, theories are divergent and use the available information somewhat selectively and in particular combinations. The next chapters can be perceived as an attempt to cross-check current concepts by making use of all the available information starting from accidental brain lesions to brain imaging studies, from animals to humans, and from normal manifestations to abnormal forms of aggression. The ultimate aim is to map the possibility of building a comprehensive neural theory of aggression control and of identifying brain alterations that result in dysfunctional aggression. This ambitious scheme necessitated the introduction of some limitations. First, the analysis was narrowed to four brain regions, namely the prefrontal cortex, amygdala, hypothalamus, and periaqueductal gray. While the number of brain regions involved in aggression is considerably larger, current views suggest that these four have critical roles. Second, overall effects attributed to certain neurotransmitters were not analyzed in detail. While neurochemical agents like serotonin may have general roles in brain function, only their local effects were found relevant for this analysis (i.e., those effects that have specific roles at particular brain locations). Third, therapeutic implications were not specifically addressed.

Scrutiny will start with the hypothalamus for two reasons: historically, this was the first brain region associated with aggression. This choice is also explainable from a neurobehavioral point of view. This is the only brain area from where aggression can be induced by stimulation in all species, and the destruction of which reduces aggression most effectively in humans.

Chapter 4
Focal Points of Aggression Control

4.1 The Hypothalamus

This brain area is a major site of vegetative control; it contains brain structures that can be considered the headquarters of the endocrine system and hosts nuclei involved in circulation, respiration, hunger, etc. The first evidence on its role in behavioral control came from aggression research. It became clear over the decades that followed this discovery that the hypothalamus controls a wide array of behavioral phenomena from sexual behaviors to feeding. It was proposed that such wide roles derive from its "strategic" position; being at the crossroads of endocrine and behavioral regulation, the hypothalamus is uniquely able to integrate these two major control mechanisms (Kruk et al. 1998). It may also play a role in the vegetative and metabolic adaptations required by the execution of behaviors, especially when these are demanding. Unfortunately, the integration of these widely different but necessarily interacting systems is poorly understood at present, for which our analysis will lack this dimension. One should keep in mind, however, that the hypothalamus likely integrates behavior with vegetative functions in a way very similar to the integration of emotions and behavior by the amygdala, and cognitions and behavior by the prefrontal cortex.

4.1.1 Hypothalamically Controlled Behaviors

Research performed in the 1920s and 1930s of the last century showed that cats and rats electrically stimulated in their hypothalamus show within seconds a variety of behaviors, e.g., aggression against conspecifics, circling, digging, drinking, escape jumps, feeding, self-grooming, hoarding food, locomotion, predation, social grooming, and teeth chattering. Subsequent research employing more sophisticated techniques delimited the hypothalamic areas subserving particular behaviors (see Table 4.1 for data in rats).

J. Haller, *Neurobiological Bases of Abnormal Aggression and Violent Behaviour*,
DOI 10.1007/978-3-7091-1268-7_4,

Table 4.1 Behaviors elicited by hypothalamic stimulation in the rat and their locus of control

Behavior	The extension of the area[a]	Nuclei involved	References
Food hoarding		LH	Herberg and Blundell (1967)
Feeding (*its inhibition*)		LH *(AHA, VMH)*	Hoebel and Teitelbaum (1962), Beltt and Keesey (1975)
Drinking		LH	Mendelson (1970)
Predatory-like attack		LH	Woodworth (1971)
Grooming	**AP**: −1−−3.25[b]; **DV**: 4.5–2; **L**: 0–1	AHA, DMH, POM, PVN, VMH,	Lammers et al. (1987)
Digging	**AP**: −1−−3.25[b]; **DV**: 4–2.5; **L**: 0.75–2.2	LH, POL, ZI	
Circling	**AP**: −1−−3.25[b]; **DV**: 4.5–2; **L**: 1.5–2.2	CAI, GP, LH, SM,	
Locomotion[b]	**AP**: −1−−3.25[b]; **DV**: 4.5–2; **L**: 0–2.2	AHA, DMH, LH, POM, POL, PVN, VMH, ZI	Lammers et al. (1988)
Escape jumps	**AP**: −1−−3.25[b]; **DV**: 4.5–2; **L**: 0–1.75	AHA, BST, DMH, FMT, LH, POM, VMH, ZI	
Rivalry attack	**AP**: −1.45−−2.95; **DV**: 3–2; **L**: 0–2	AHA, LH, VMH	Lammers et al. (1988)
Social grooming	**AP**: −1−−3.25[b]; **DV**: 4–2; **L**: 1–2	LH, POL, VMH	
Teeth chattering	**AP**: −1−−2.95; **DV**: 3.5–1; **L**: 0–1.5	AHA, POM, VMH	

Note that in addition to the "nuclei involved" extended neuron-sparse regions ("white matter") between and around the listed nuclei also induced responses to stimulation. The thoroughness of the studies was widely different. Therefore, the available information varies from study to study. Note that overlaps in the "nuclei involved" may or may not indicate real overlaps. In many cases, different parts of the same nucleus were involved in different behaviors. For such subtle differences, see the cited papers

Abbreviations: *AP, DV, and L* anteroposterior, dorsoventral, and lateral from Bregma, *AHA* anterior hypothalamic area, *BST* bed nucleus stria terminalis, *CAI* capsula interna, *DMH* dorsomedial hypothalamic nucleus, *FMT* fasciculus mamillothalamicus, *GP* globus pallidus, *LH* lateral hypothalamic area, *POL* nucleus preopticus lateralis, *POM* nucleus preopticus medialis, *PVN* paraventricular hypothalamic nucleus, *SM* stria medullaris thalami, *VMH* ventromedial hypothalamus, *ZI* zona incerta

[a]The regions extended beyond these limits in some cases but no other areas were investigated

[b]Walking crisscross in the test cage

These studies led to the conclusion that positive sites for any of these behavioral responses are clustered into restricted and only partially overlapping hypothalamic areas. However, neuronal clusters with specific behavioral roles have several interesting properties that all raise questions regarding the nature of the neural mechanisms activated:

1. The anatomical boundaries of hypothalamic nuclei—as identified by neuron clustering in brain atlases are not "respected" by functional boundaries as identified by electrical stimulation. For example, particular behaviors can be elicited from relatively large areas that comprise several "anatomical nuclei" and from neuron-sparse locations. The latter are sometimes rather distant from the surrounding (neuron-dense) nuclei from which the same behavior cannot always be elicited, suggesting that cell-scarce areas play important roles in behavioral control. Thus, both electrophysiological and anatomical techniques provide hypothalamic "maps," but neuronal and functional clustering is different.
2. The hypothalamic "electrophysiological nuclei" overlap spatially, as the stimulation of the very same nucleus may activate different behaviors. Yet these spatially overlapping areas often appear functionally different. For example, teeth chattering, escape jumps, and attacks can be elicited from overlapping hypothalamic subdivisions but at different current intensities, suggesting either that different neuronal clusters are activated at different current intensities or that the same neuronal cluster elicits different responses depending on the intensity of stimulation.
3. Hypothalamic stimulation elicits behaviors only when the targets of those behaviors are present. For example, hoarding behavior occurs only when there is food to carry; feeding centers do not induce faked feeding if food is not available; aggression is evoked only when a rat is nearby, etc. Stimulation usually leads to no behavioral response if the target is missing.
4. Last but not least, the stimulation of hypothalamic sites evokes very complex behavioral patterns. For example, when hoarding-related hypothalamic areas are stimulated, rats scan the environment, locate food, approach it, seize with teeth, and hoard it to a particular location within the cage; this behavior is repeated till all the food pellets are gathered at one and the same location. At the same time, certain elements of naturally occurring behaviors are missing when the same behavior is elicited by hypothalamic stimulation. For example, teeth chattering (a threat-like behavior) and attacks are elicited from partly overlapping but distinct hypothalamic sites, while other components of the aggressive repertoire (e.g., lateral threat) cannot be induced by hypothalamic stimulation.

The interpretation of the behavioral consequences of hypothalamic stimulation raised many controversies. Initially it was suggested that the effects of stimulation are largely mediated by passing fibers, e.g., those of the mediobasal bundle (Ranck 1975; Velley et al. 1988; Gratton and Wise 1988). It was shown, however, that hypothalamically evoked locomotor stepping persists when passing fibers are lesioned (Sinnamon 1990), but are inhibited by agents that locally inhibit synaptic transmission (López and Carrer 1985); moreover, behaviors elicited by electrical stimulation can also be induced in the very same regions by neurochemical agents that either stimulate (e.g., glutamate agonists) or disinhibit (e.g., GABA antagonists) neuronal activity (*aggression*: Adams et al. 1993; Haller et al. 1998a, b, c; Roeling et al. 1993; *grooming*: Roeling et al. 1990; *locomotor stepping*: Marciello and Sinnamon; 1990; *feeding*: Dube et al. 1999). In addition, it was recently shown that aggression can be induced from the hypothalamus by optogenetic stimulation (Lin et al. 2011). Thus, hypothalamically evoked behaviors result from local neuronal activation that propagates to a large number of brain areas involved in behavioral control by the extensive efferent connections of hypothalamic neurons as recognized very early (Anschel et al. 1982).

It was also suggested that hypothalamic stimulation affects motivational processes that result in context-dependent behavioral responses (Berridge and Valenstein 1991; Valenstein et al. 1968). This hypothesis is, however, invalidated by several arguments. For example, locomotor stepping could be elicited in anesthetized rats by glutamate injections into the lateral hypothalamus, which excludes both the "passing fiber" and the "motivational" hypothesis (Marciello and Sinnamon 1990). As it regards aggression, hypothalamic stimulation can separately elicit certain elements of aggressive behavior (e.g., teeth chattering and biting attacks), while offensive postures (e.g., lateral threat, upright boxing) and dominance are never seen in response to hypothalamic stimulation. One can argue that if stimulation increased the motivation for aggression, then the response should have included behaviors related to that particular motivational state.

Taken together, these findings demonstrate that the hypothalamus comprises neural mechanisms ("electrophysiological nuclei") from where complex and context-dependent behaviors can specifically be evoked by electric stimulation. The hypothalamus is unique in that this is the only brain area in the rat from where biting attacks can be elicited by various means including electrical stimulation, glutamate receptor agonists, GABAA receptor antagonists, and optogenetic stimulation.

4.1.2 Hypothalamic Control of Aggression in Animals and Humans

4.1.2.1 Anatomical Aspects

This line of research was initiated by the seminal work of Hess performed in rats and cats (1928). Subsequent studies showed that the phenomenon is general: biting attacks can be elicited by hypothalamic stimulation in virtually all the species studied so far, suggesting that this mechanism of aggression control is evolutionarily conserved (e.g., *cats*: Romaniuk 1965; *chicken*: Putkonen 1966; *cichlid fish*: Demski 1973; *guinea pig*: Martin 1976; *humans*: Bejjani et al. 2002; *lizards*: Sugerman and Demski 1978; *monkeys*: Lipp 1978; *mice*: Lin et al. 2011; *opossum*: Roberts et al. 1967; *rats*: Vergnes and Karli 1969; Kruk et al. 1979). In lizards at one extreme, aggression can be induced from several brain areas while at the other

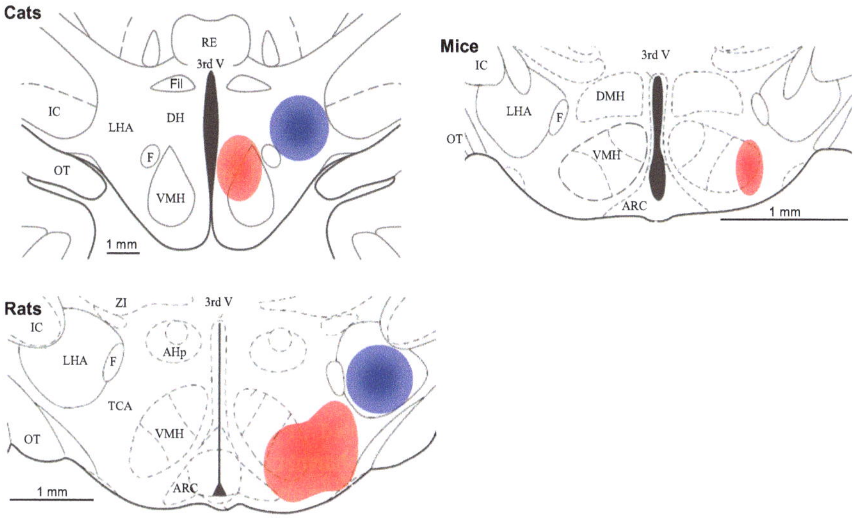

Fig. 4.1 *The anatomical localization of hypothalamic attack sites in three species. Red*, affective/rivalry aggression; *blue*: predatory attacks (not known in mice); *3rd V* 3rd ventricle; *AHp* anterior hypothalamic nucleus, posterior part; *ARC* arcuate nucleus; *DH* dorsal hypothalamic area; *DMH* dorsomedial hypothalamic nucleus; *F* fornix; *Fil* nucleus filiformis; *IC* internal capsule; *LHA* lateral hypothalamic area; *OT* optic tract; *PVN* paraventricular hypothalamic nucleus; *RE* nucleus reuniens; *VMH* ventromedial hypothalamic nucleus; *TCA* tuber cinereum; *ZI* zona incerta. Based on Siegel et al. (1999), Lammers et al. (1988) and Lin et al. (2011)

extreme the hypothalamus is the only area from where attacks can reliably be elicited in rats (Kruk 1991). Hypothalamic structures subserving attack were best described in cats and rats, and substantial evidence is available in mice, hamsters, and humans; therefore, we will focus on these species.

The stimulation of the hypothalamus evokes two types of aggression in cats, namely defensive rage and predatory attacks. The former consists of threatening vocalizations and postures and strikes upon provocation. Predatory attack also called quiet biting involves the killing of pray (e.g., a rat) by biting the back of its neck without vocalizations or threat postures (Siegel et al. 1999). Hypothalamic structures controlling defensive rage and predatory aggression are topographically and functionally separated: the relevant hypothalamic structures (ventral/medial and lateral to the fornix, respectively) are not only behaviorally specific, but the two areas also inhibit each other (Cheu and Siegel 1998) (Fig. 4.1). Topographic organization is also revealed by the distinct neural connections and pharmacology of the two regions (Brutus and Siegel 1989; Shaikh et al. 1984; Siegel et al. 1999, 2007). The interpretation of hypothalamically evoked behaviors needs some additional interpretation especially in conjunction with behaviors elicited from analogous hypothalamic structures of rodents. "Rage" induced from ventral/medial areas of the hypothalamus is usually considered a defensive response, the main component of which is hissing that is executed by, and in parallel with, the opening of the jaws. This interpretation involves that the cat hypothalamus contains no mechanisms for attacks on conspecifics, in sharp contrast to all other mammals where such attacks can readily be evoked. We believe, however, that this strange conclusion derives from the specific conditions under which this type of studies is performed. In rats, stimulation is always performed in the presence of a potential opponent; in cats, this element is missing probably because of ethical reasons; carnivores can inflict heavy damage on opponents. Nevertheless, one can reasonably assume that attacks were performed if an opponent was present, and the jaws opened for hissing were sooner or later closed for biting. This is supported by the behavioral similarity of the initial phases of intraspecific contests and "defensive rage." Consequently we assume that areas involved in "defensive rage" are in fact components of the system that controls "affective aggression." In the following we will prefer the latter over the former phrase.

The study of hypothalamic structures involved in attack followed a different trajectory in the rat. The precise location and pharmacological responsiveness of the mediobasal hypothalamus (called "hypothalamic attack area") was described in detail, while brain areas involved in predatory aggression received less attention. Similar to cats, attacks on conspecifics can be induced in rats by the stimulation of the mediobasal hypothalamus, which is located ventral and medial to the fornix and includes the posterior part of the anterior hypothalamus, the relatively cell-sparse region extending from the fornix to the third ventricle, and the ventrolateral aspects of the ventromedial hypothalamus (Kruk et al. 1979; Kruk 1991; Lammers et al. 1988). Later studies demonstrated that spontaneous aggression induces c-Fos activation in the very same region (Halasz et al. 2002a); moreover, the unilateral electric stimulation of the mediobasal hypothalamus induces attacks only when the contralateral side is also activated in a manner, and to the extent, similar to that seen in spontaneous aggression (Halasz et al. 2002b). The presence of predatory attack sites was also demonstrated in rats. For example, the electrical and neurochemical stimulation of the lateral hypothalamic area induces frog and/or mouse killing (Vergnes and Karli 1969; Smith et al. 1970; Bandler 1970; Woodworth 1971), and recent findings from our laboratory show that the lateral hypothalamus is strongly activated by mouse killing (Tulogdi et al. 2014). However, the topographic organization of the two types of aggression may not be as distinct as in cats. Some of the early studies revealed that the stimulation of the lateral hypothalamus does induce attacks on conspecifics albeit this response required considerably larger current intensities than muricide (Woodworth 1971). Although no attacks on conspecifics were induced from the lateral hypothalamus in other labs (Kruk 1991; Siegel et al. 1999), it was shown that the stimulation of the lateral hypothalamus promotes spontaneous attacks under certain circumstances (Koolhaas 1978). The assumption that the lateral hypothalamus is involved in intraspecific aggression is also supported by c-Fos and brain imaging studies showing that the lateral hypothalamic area is activated during resident-intruder conflicts (Clinton et al. 2011; Ferris et al. 2008; Tulogdi et al. 2010). Taken together, these findings show that intraspecific aggression and predatory aggression are controlled by hypothalamic areas medial and lateral to the fornix, respectively, but the two mechanisms' area not as distinct as in the cat.

It was recently demonstrated that the optogenetic stimulation of the lateral aspects of the ventromedial nucleus rapidly induces coordinated and directed biting in mice (Lin et al. 2011). Although bites were elicited from a very restricted hypothalamic area in the mouse, a wider array of hypothalamic mechanisms are activated by territorial aggression in this species, suggesting that many hypothalamic structures analogous to the rat hypothalamic attack area are also involved in mouse aggression (e.g., the posterior part of the anterior hypothalamus, the ventrolateral nucleus of the hypothalamus, and the tuber cinereum; Duncan et al. 2009; Haller et al. 2006). It was also shown that individual variation in estrogen receptor α immunoreactivity in the anterior hypothalamus (part of "hypothalamic attack area" in rats) is positively correlated with aggressive behavior in mice (Trainor et al. 2006). Moreover, some findings suggest that the lateral hypothalamus also plays a role in mouse aggression (Duncan et al. 2009; Hasen and Gammie 2006). Taken together, these findings suggest that mouse aggression is controlled by hypothalamic structures analogous to those controlling aggression in rats.

In hamsters, the location and extent of hypothalamic structures involved in aggressiveness were not precisely delimited, but it is believed that the key regions are the anterior and lateral hypothalamic areas which control intraspecific attacks (Ferris and Delville 1994; Knyshevski et al. 2005).

Evidence on the involvement of the human hypothalamus in aggression came from the study by Sano et al. (1970) who lesioned posteromedial regions of the hypothalamus, which reduced or abolished aggressiveness in violent patients (Fig. 4.2). The lesioned region included medial parts of the hypothalamus and the medial subdivision of the lateral hypothalamus. These findings were replicated by many other studies (Dieckmann et al. 1988; Laitinen 2001; Pedrosa-Sanchez and Sola 2003; Ramamurthi 1988). A case study suggested that the electrical stimulation of the very same area induces aggressiveness (Bejjani et al. 2002). The behavioral response was similar to that seen in animals as it occurred within seconds, and had no post-stimulation effects on aggression. Importantly, the patient although clearly remembered it was not able to explain his/her behavior, suggesting that the hypothalamically evoked aggressiveness had no obvious cognitive aspect. The role of the human hypothalamus in aggression control was supported by recent studies showing that the continuous stimulation of the "triangle of Sano" for month and years suppresses aggression in patients with intractable aggression (Franzini et al. 2005, 2010; Hernando et al. 2008). At the first sight, reduced violence obtained by the Sano and the Franzini approaches is intriguing, as lesions and stimulations were targeted to very similar hypothalamic sites. It is believed, however, that deep brain stimulation is in fact a functional lesion (Bennazzous and Hallett 2000); high-frequency stimulation can rapidly render neurons inactive by a phenomenon called stimulation-induced inhibition of depolarization. High-frequency stimulations may even lesion the area if performed over prolonged periods. Thus, the Sano and Franzini procedures are very similar in their ultimate neural outcome. An interesting example of hypothalamic aggression control is the

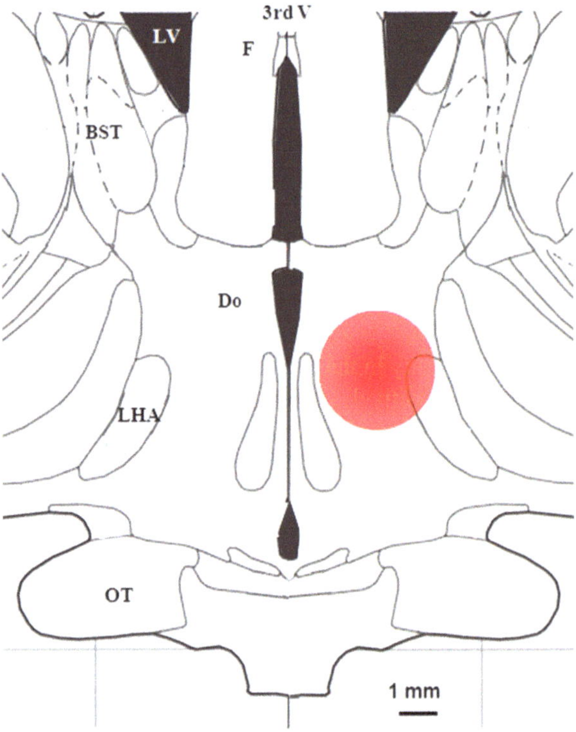

Fig. 4.2 *The anatomical localization of hypothalamic lesions that reduce aggression in severely violent patients. 3rd V* 3rd ventricle; *BST* bed nucleus of stria terminalis; *Do* dorsal hypothalamic nucleus; *LHA* lateral hypothalamic area; *LV* lateral ventricle; *F* fornix; *OT* optic tract. Note the marked similarity in location and size with the areas from where aggression can be induced by electrical stimulation (Fig. 4.1)

hypothalamic hamartoma. A subgroup of patients developing this malformation show increased aggression, while the removal of the affected hypothalamic parts abolishes aggressiveness (de Almeida et al. 2008; Weissenberger et al. 2001). The impact of hamartomas on hypothalamic function is poorly known, but the findings with this malformation show that hypothalamic functions are tightly linked to aggression in humans. This assumption is further supported by a brain imaging study performed in females exposed to threatening stimuli; a significant activation of the hypothalamus was noticed (Hermans et al. 2008).

Taken together, these findings suggest that there is a region in the human hypothalamus that controls aggressiveness, and disparate findings suggest that this area responds to stimulation in a manner similar to that seen in laboratory animals.

4.1.2.2 Neurochemical Aspects

In the cat, the affective aggression-bound medial hypothalamus is stimulated from upstream by substance P-ergic medial amygdala neurons; its "execution-bound" downstream outputs to the periaqueductal gray are glutamatergic (Siegel et al. 1999, 2007). Inhibitory outputs to the lateral hypothalamus are GABAergic.

The mediobasal hypothalamus also receives facilitatory noradrenergic, dopaminergic, and cholinergic inputs and inhibitory serotonergic inputs. The role of serotonin appears complex as effects exerted via different receptors (e.g., 5-HT_{1A} and 5HT_{2C}) affect aggression in an opposite fashion (Hassanain et al. 2003; Siegel et al. 1999). The neurochemical nature of lateral hypothalamic neurons is known in less detail.

The rat mediobasal hypothalamus (involved in attacks on conspecifics) contains both glutamatergic and GABAergic neurons that have a distinct localization within the area (Hrabovszky et al. 2005). The same study showed that a subgroup of glutamatergic neurons also expresses thyrotropin-releasing hormone, the role of which remains unclear at present. From upstream, the system is activated by substance P neurons located in the medial amygdala, a mechanism that seems very similar to that seen in cats (Halasz et al. 2008, 2009). Local glutamatergic neurons are responsible for the induction of attacks, whereas GABAergic mechanisms locally inhibit them (Adams et al. 1993; Haller et al. 1998a, b, c; Roeling et al. 1993). Hypothalamic attack is under the control of several other factors, namely serotonin (local inhibition; Kruk 1991) and vasopressin (stimulation; Caffrey et al. 2010; Ferris et al. 2008). The latter mechanism is sensitive to testosterone (Trainor et al. 2006).

In hamsters, the anterior hypothalamus (i.e., the "hypothalamic attack area" in this species) is under the control of vasopressinergic inputs (Albers et al. 2006; Ferris and Delville 1994; Ferris and Potegal 1988) that seem to integrate other neurochemical and hormonal influences, namely those of serotonin and testosterone (Delville et al. 1996; Ferris et al. 1997). Attack-related hypothalamic mechanisms are also sensitive to the local application of glucocorticoids (Hayden-Hixson and Ferris 1991a, b).

Findings on the hypothalamic neurochemistry of aggression are sparse in humans. Nevertheless, a study in females suggests that the testosterone/cortisol ratio has a role (Hermans et al. 2008); a role for serotonin was also shown (Koch et al. 2007).

4.1.2.3 Overview

The hypothalamus hosts cell groups that are critically involved in the induction of attack behavior. Two regions of interest were identified, each being involved in a specific form of attack. The medial parts of the hypothalamus induce attacks on conspecifics, while lateral parts induce attacks on prey. While the two mechanisms are distinct in cats, the lateral hypothalamus seems to have a somewhat mixed role in rodents. Hypothalamic mechanisms are also involved in the control of human aggression.

Hypothalamic mechanisms appear to be controlled by substance P from upstream (particularly from the medial amygdala) and send glutamatergic projections downstream. The functioning of attack-bound hypothalamic mechanisms is controlled by inputs that utilize acetylcholine, dopamine, GABA, noradrenaline, serotonin, and vasopressin as neurotransmitters. Notably, these are the neurochemical mechanisms that are believed to control aggression overall; the studies reviewed above suggest that they have local impacts at the level of aggression-bound hypothalamic structures. Testosterone and glucocorticoids also have a role at this level. Neurochemical and neuroendocrine data in humans are sparse, but the available findings are in line with those obtained in animals.

4.1.3 Hypothalamic Mechanisms of Abnormal Aggression in Animals[1]

4.1.3.1 Overall Activation of the Hypothalamus

One approach to the study of brain control is the elucidation of overall activities seen in particular brain areas. "Overall" means here that neuron types, neurochemical agents, receptors, etc., are not distinguished; the technique provides indistinct information on the level of activation. This is one of the main approaches in human brain imaging techniques; their rodent analogs are c-Fos studies or other studies of this kind.[2]

One may have expected that hypothalamic structures involved in attack were overactivated in abnormal aggression if the "hyperactivation theory" was right, i.e., if abnormal aggression resulted from the exacerbated activation of aggression-controlling brain mechanisms. However, hypothalamic mechanisms controlling attack undergo interesting, sometimes surprising, changes in rodent models of abnormal aggression.

In some models, the expected changes were seen. The mediobasal hypothalamus was overactivated by aggression in four models, namely in the postweaning social isolation, anabolic steroid, and cocaine models, as well as in rats selected for low anxiety (Toth et al. 2012; Ricci et al. 2007; Knyshevski et al. 2005; Beiderbeck et al. 2012). As it regards the lateral hypothalamus (implicated mainly in predatory attacks), increased activation was seen in three models (the glucocorticoid dysfunction and cocaine model, as well as in low-anxiety rats) (Beiderbeck et al. 2012; Knyshevski et al. 2005; Tulogdi et al. 2010).

In other models, however, no changes in mediobasal hypothalamic activation were seen, but the lateral hypothalamus was activated where this area was investigated. Such findings were observed in the glucocorticoid dysfunction model, in mice selected for high aggressiveness, and in rats selected for high anxiety

[1] No data in cats are available, because all models of abnormal aggression utilize rodents.

[2] C-Fos belongs to a class of genes that are summarized under the generic term "immediate early genes." Such genes are rapidly activated in neurons that are particularly active (e.g., are activated by a behavioral test). Their end product (e.g., the c-Fos protein) can be labeled immunohistochemically and can serve to identify neurons that were active during the behavioral test. This is the basis of identifying neuronal activation patterns associated with behaviors.

(Beiderbeck et al. 2012; Haller et al. 2006; Halasz et al. 2002a; Tulogdi et al. 2010). Lateral hypothalamic activation was not increased in rats selected for high anxiety and those submitted to postweaning social isolation (Beiderbeck et al. 2012; Toth et al. 2012). In rats selected for high anxiety, neither the mediobasal nor the lateral hypothalamus was overactivated despite clearly abnormal attack patterns (Beiderbeck et al. 2012).

Interestingly, this pattern of activity and nonactivity overlaps with the emotional responses associated with these models of abnormal aggression. The mediobasal hypothalamus was overactivated mainly in models where aggression is associated with hyperarousal, e.g., the postweaning social isolation and cocaine models both being characterized by increased glucocorticoid and autonomic stress responses (Ansah et al. 1996; Moldow and Fischman 1987; Toth et al. 2011). In contrast, the mediobasal hypothalamus was not activated in models characterized either by low glucocorticoid and autonomic stress responses or by low basal glucocorticoid levels and heart rates i.e., by hypoarousal (*rat*: Haller et al. 2001, 2004; *mice*: Caramaschi et al. 2008a, b; Veenema et al. 2004). Although aggression-induced autonomic activation was not directly investigated in rats selected for anxiety, studies on glucocorticoids support the assumption that hypo- and hyperarousal during aggressive confrontations have an impact on the activation of hypothalamic centers involved in attack. In highly anxious rats that attacked vulnerable targets, the glucocorticoid response to resident-intruder conflicts was decreased and no aggression-induced hypothalamic activation was noticed (Neumann et al. 2010). In the low-anxiety strain that showed vulnerable attacks and also attacked females and anesthetized opponents, the glucocorticoid response to resident-intruder conflicts increased, and the hypothalamus was strongly activated by fights.

The findings briefly reviewed above suggest that hypoarousal-associated aggression does not, while hyperarousal-associated aggression does correlate with increased mediobasal hypothalamic activation. This relationship is illustrated in Fig. 4.3 by the comparison of the glucocorticoid deficiency postweaning social isolation models. The direct comparability of these models is ensured by the consistency of behavioral and neuroanatomical analysis, as the findings were obtained by the same group. In the hypoarousal model, normal attack counts but abnormal attack features are associated with a "predatory-like" hypothalamic activation pattern (normal mediobasal but increased lateral hypothalamic activation; Tulogdi et al. 2010). In contrast, exacerbated "rivalry-like" hypothalamic activation pattern was seen in the postweaning social isolation (hyperarousal) model (increased mediobasal but normal lateral hypothalamic activation; Toth et al. 2012). Unfortunately, evidence on predation-related hypothalamic control mechanisms is lacking in mice, while the arousal component of aggression is poorly known in other models. Nevertheless, the findings briefly reviewed above and the example summarized in Fig. 4.3 suggests that the emotional component of abnormal attacks has a large impact on the hypothalamic control of aggression.

Glucocorticoid deficiency model		Post-weaning social isolation model
decreased	*Arousal*	increased
Yes	*Attacks on vulnerable targets*	Yes
Yes	*Deficient attack signaling*	Yes
No	*Increased attack counts*	Yes
No	*Offensive ambiguity*	Yes
No	*Behavioral agitation*	Yes

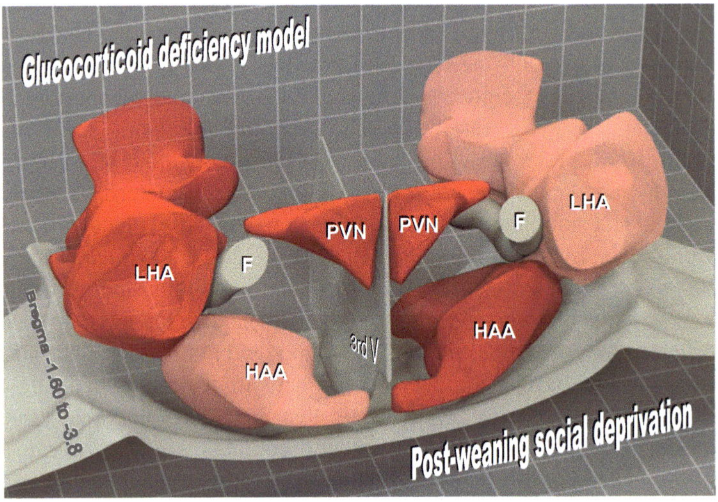

Fig. 4.3 *Associations between emotionality and aggression-induced hypothalamic activation patterns* as illustrated by the comparison of the glucocorticoid deficiency and post-weaning social isolation models (Tulogdi et al. 2010; Toth et al. 2012). The models are directly comparable as subjects were rats in both cases, and behavioral and neuroanatomical analysis was highly consistent (the findings were obtained by the same group). *Upper panel*: emotional and behavioral characteristics in the two models. *Lower panel*: 3D reconstruction of the relevant hypothalamic region. *Pink* activation similar to controls; *red* over-activation compared to controls; *3rdV* 3rd ventricle; *F* fornix; *HAA* hypothalamic attack area (mediobasal hypothalamus); *LHA* lateral hypothalamic area; *PVN* paraventricular hypothalamic nucleus

4.1.3.2 Serotonin

The "serotonin hypothesis" is considered a cornerstone of aggression control (Nelson and Chiavegatto 2001). Theoretically, abnormal aggression may be driven by decreased serotonergic control, which would indirectly upregulate the function of brain mechanisms that promote aggression. However, the role of hypothalamic serotonin is unexpectedly complex in models of abnormal aggression.

In the mouse variant of the postweaning social isolation model, rats submitted to maternal separation as well as anabolic steroid-treated hamsters, findings support the notion that serotonin suppresses aggression. In these models, serotonin fiber density and/or the density of various serotonin receptors decreased in the mediobasal hypothalamus that is involved in the control of rivalry attacks (Bibancos et al. 2007; Grimes and Melloni 2002, 2005; Ricci et al. 2006; Veenema et al. 2006). Some contradictions do exist; e.g., $5HT_{1A}$ receptor density decreased in hamsters treated with anabolic steroids, but remained unchanged in mice submitted to postweaning social isolation (Bibancos et al. 2007; Ricci et al. 2006). Overall, however, a decrease in hypothalamic serotonin functions is evident in these models.

In sharp contrast with the aforementioned models, hypothalamic $5\text{-}HT_{1B}$ and $5\text{-}HT_{2A}$ receptor mRNA levels doubled in mice submitted to the alcohol model (Chiavegatto et al. 2010). Hypothalamic $5\text{-}HT_6$ receptor expression also doubled in the anabolic steroid model (Ambar and Chiavegatto 2009), while the density of serotonin fibers and the number of serotonergic varicosities increased in the anterior hypothalamus of hamsters submitted to the subjugation model (Delville et al. 1998; Ferris and Potegal 1988). Similar findings were obtained with N-type Ca channel $alpha_{1B}$ subunit-KO mice where attack counts increased over five times, yet the serotonin content of the hypothalamus doubled without parallel increases in hypothalamic noradrenaline and dopamine (Kim et al. 2009).

One has to conclude that while hypothalamic serotonin undergoes the expected changes in some models, certain forms of abnormal aggression are associated with *increased* serotonergic neurotransmission in the hypothalamus. Findings with the postweaning social isolation and alcohol models are especially intriguing; as the subjects were mice in both cases, the findings were obtained by an overlapping set of authors, and the immunocytochemical methodologies were highly similar (Fig. 4.4). One can argue that the intriguing changes seen at hypothalamic level were "overwritten" by reduced serotonergic neurotransmission in other brain areas, e.g., the prefrontal cortex (Chiavegatto et al. 2010). However, this argument does not hold true for two reasons. First, increased serotonergic neurotransmission in abnormally aggressive animals was seen in extra-hypothalamic sites as well. For example, Delville et al. (1998) observed increased number of septal 5-HT varicosities in the hamster "subjugation" model. Chiavegatto et al. (2010) showed that in addition to the hypothalamus, the expression of $5\text{-}HT_{1B}$ receptors was increased in the amygdala of mice that showed excessive increases in aggression after alcohol, while Bibancos et al. (2007) reported increased $5\text{-}HT_6$ receptor expression in the hippocampus of mice submitted to the postweaning isolation model. Second, paradoxical associations were also noticed when neurotransmitter/behavior correlations were investigated. There was no significant correlation between raphe serotonergic activation and aggressiveness in the hypoarousal model which was in sharp contrast with data obtained in controls (Haller et al. 2005a, b). In addition, the $5\text{-}HT_{1A}$ partial agonist buspirone ameliorated normal (rivalry) aggression but dramatically increased attack counts in the glucocorticoid deficiency model (Haller et al. 2007). In a similar fashion, there was no obvious link between serotonergic neurotransmission and aggressiveness in mice selected for aggressiveness (Natarajan et al. 2009a, b, c; Wallinga et al. 2009). Moreover, $5\text{-}HT_{1A}$ and $5\text{-}HT_{1B}$ agonists decreased aggressiveness in highly aggressive feral rats by decreasing serotonin release via their presynaptic autoreceptor function (de Boer and Koolhaas 2005).

Taken together, these findings demonstrate that the anti-aggressive effects of serotonergic neurotransmission are compromised in certain models of abnormal aggression; moreover, serotonin may paradoxically promote aggressiveness in

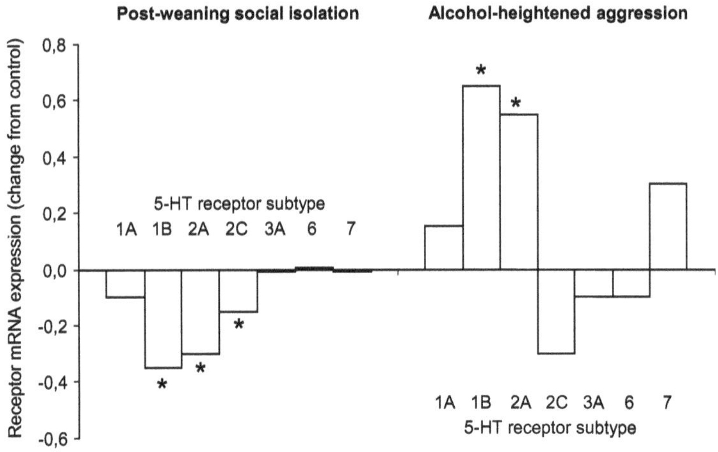

Fig. 4.4 *Serotonin neurotransmission often shows opposite changes in different models of abnormal aggression* as illustrated by a comparison between the post-weaning social isolation and the alcohol model of abnormal aggression. The two models are directly comparable as the subjects were mice in both cases, the immunocytochemical methodologies were highly similar, and the findings were obtained by an overlapping set of authors. The figure was compiled from data published by Bibancos et al. (2007) and Chiavegatto et al. (2010). The *Y*-axis shows the difference between transcript amounts obtained in the experimental groups and their controls. *Significantly different from controls in the respective publications

some models. Most importantly, these paradoxical interactions were seen in models where abnormal aggression was associated with hypoarousal (*blunted cortisol response to aggression in subjugated hamsters*: Ferris et al. 2005; *low basal corticosterone after adolescent anabolic steroid treatment*: Rejeski et al. 1990; *low corticosterone stress responses and low basal heart rate in mice selected for high aggressiveness*: Caramaschi et al. 2008a, b; Veenema et al. 2004; *low basal CORT and low autonomic responses to aggression in the rat glucocorticoid dysfunction model*: Haller et al. 2001, 2004).

4.1.3.3 Vasopressin

Vasopressin plays important roles in aggression control (see above); as such, one can hypothesize that increased vasopressin neurotransmission promotes abnormal forms of aggression. This, however, is not always the case.

Vasopressin neurotransmission appears to promote abnormal manifestations of aggression in several models. The interaction has many facets. In the anabolic steroid model, the expression of the V_{1A} vasopressin receptor increased in the ventrolateral but not in the anterior hypothalamus (DeLeon et al. 2002). Despite this, V_{1A} antagonists administered into the anterior hypothalamus decreased the anabolic steroid-induced aggression (Harrison et al. 2000). In the amphetamine model, V_{1A} expression increased in the anterior hypothalamus, and V_{1A} antagonists abolished amphetamine-induced aggression (Gobrogge et al. 2009). V_{1A} antagonists also abolished exacerbated aggression in the cocaine model, where an increased stimulation-induced vasopressin release was noticed without changes in vasopressin content or vasopressin fiber density (Jackson et al. 2005). In the maternal separation model (the only rat study addressing hypothalamic vasopressin), no change in vasopressin expression was noticed in the anterior hypothalamus, but vasopressin expression (both basal and aggression induced) increased in the lateral hypothalamus (Veenema et al. 2006). Despite this variability of findings, increased vasopressin functions are evident in these models.

Changes seen in other models were opposite to expectations. Decreased anterior hypothalamic vasopressin content and decreased vasopressin fiber density was noticed in the subjugation model (Delville et al. 2000). Gene polymorphism studies suggest similar changes in rats selected for anxiety. Low-anxiety rats express a vasopressin gene that has a reduced transcription rate, and consequently vasopressin production is lower in this than in the high-anxiety strain (Bunck et al. 2009; Kessler et al. 2007). The low-anxiety strain showed more disturbed attack patterns than the high-anxiety strain; consequently, higher abnormality of aggression was associated with a lower hypothalamic vasopressin levels.

Similar to overall activation levels and serotonin neurotransmission, the role of hypothalamic vasopressin in abnormal aggression seems model dependent. While it clearly promotes abnormal aggression in the anabolic steroid, amphetamine, cocaine, and maternal separation models, the opposite is true for the subjugation model and rats selected for low-anxiety behavior.

4.1.3.4 Other Neurotransmitters

There are neurotransmitter systems that were studied in one or a few models only. This does not diminish the importance of the findings, or the relevance of the respective systems for abnormal aggression, but the various models of abnormal aggression cannot be compared in their case. For example, the impact of substance P neurotransmission (the effects of which are mediated by neurokinin 1 (NK_1) receptors) was investigated in the glucocorticoid dysfunction model only (Halasz et al. 2008, 2009). In this model, mediobasal hypothalamic neurons expressing NK_1 receptors were specifically overactivated by aggressive encounters; moreover, the systemic application of NK_1 blockers or the selective destruction of NK_1 receptor expressing mediobasal hypothalamic neurons by saporin-conjugated substance P dramatically reduced violent biting, without affecting milder forms of aggression. Thus, NK1-expressing mediobasal hypothalamic neurons appear specifically involved in abnormal features of aggression in this model.

In the anabolic steroid model, dopamine production by the circular and medial supraoptic nuclei of the hypothalamus was increased, and the expression of D_2 dopamine receptors was increased in the anterior hypothalamus (the attack area in the species studied; Ricci et al. 2009; Schwartzer et al. 2009). Thus, dopamine neurotransmission may be another mechanism that promotes abnormal aggression at hypothalamic level.

Finally, the role of glutamatergic and GABAergic neurotransmission was also investigated in a few models. In the hamster anabolic steroid model, glutamate synthesis increased in both the anterior and lateral hypothalamus, but equivocal changes were noticed at the level of their projections (Carrillo et al. 2009; Fischer et al. 2007). The role of GABA was investigated in the anabolic steroid model (Grimes et al. 2003; Schwartzer et al. 2009), in the cocaine model (Ricci et al. 2005), and in mice selected for high aggressiveness (Weerts et al. 1992). In the latter model, findings were in line with expectations: GABA binding was increased in low aggression genotype and was decreased in the highly aggressive genotype. In the former two models, however, increased GABA synthesis was noticed in the anterior hypothalamus despite the fact that GABA agonists inhibit rivalry aggression when infused into mediobasal hypothalamic sites (Adams et al. 1993; Roeling et al. 1993).

The above findings reveal that GABAergic mechanisms may also differentiate certain models of abnormal aggression. Comparative data are not available for substance P, dopamine, and glutamate, but the available findings show that they are also relevant from the point of view of hypothalamic mechanisms involved in abnormal aggression.

4.1.4 The Human Case

Unfortunately, human data on the role of the hypothalamus in abnormal forms of aggression are extremely scarce. It was shown in psychopaths that hypothalamic activation correlates positively with aggressiveness shown in the Taylor Aggression Paradigm and the response correlated with increased anger (Veit et al. 2010). An earlier study performed by the same set of authors and employing similar techniques showed no such a correlation in healthy controls (Lotze et al. 2007). As anger per se does not induce hypothalamic activation (Damasio et al. 2000), these findings suggest that the delivery of severe punishment correlates with increased hypothalamic activation in psychopaths but not in healthy subjects. A very recent publication showed that gray matter was increased in the hypothalamus of borderline personality patients, and this increase correlated positively with the history of traumatization (Kuhlmann et al. 2013). As symptomatology included high levels of aggression among the subjects of this study, this finding tentatively suggests that traumatic experience results in increased hypothalamic gray matter, which contributes to the expression of aggressiveness. In another study, borderline personality patients showed decreased serotonergic functions in the hypothalamus (Koch et al. 2007).

The above findings are consistent with the assumption that the hypothalamus contributes to the expression of abnormal aggression in humans, while serotonergic neurotransmission is a negative modulator of such interactions. In alcoholic people charged with domestic violence, however, glucose metabolism *decreased* in the hypothalamus (George et al. 2004). While interactions with alcoholism cannot be ruled out, no similar change was obtained in nonviolent alcoholics.

The scarcity of data precludes firm conclusions on the role of the hypothalamus in aggression-related psychopathologies in humans. In addition, the spatial resolution of current brain imaging techniques does not distinguish the various subregions of the hypothalamus, which have rather different roles in brain function and behavior. Tentatively, however, data suggest that aggression-related psychopathologies are associated with increased hypothalamic gray matter volume, increased hypothalamic activation during retaliation, and decreased serotonergic control at the level of the hypothalamus. As the aggressiveness of psychopaths correlated with anger that was induced by the experimental punishment received, while borderline patients show signs of hyperarousal in terms of both glucocorticoid production and autonomic activation (Carvalho et al. 2012; Lyons-Ruth et al. 2011; Weinberg et al. 2009), the above inference seems to be valid to emotional/reactive (hyperarousal-driven) aggression.

One study found a different kind of interaction suggesting that the role of the hypothalamus in abnormal human aggression replicates the diversity seen in animals.

4.1.5 Overview: Etiological Factor-Dependent Alterations in the Hypothalamus

At the first sight, models of abnormal aggression provided conflicting data regarding the role of the hypothalamus. However, discrepancies disappeared when the emotional background of subjects was taken into account (Table 4.2[3]). This grouping and limited data on the lateral hypothalamus suggest that abnormal aggression is associated with two basic patterns of aggression-induced hypothalamic activation:

1. When abnormal aggression is performed on the background of increased arousal, the overall activation of the mediobasal hypothalamus increases, the serotonergic control of the hypothalamus diminishes, and hypothalamic vasopressinergic neurotransmission increases.
2. In aggression models associated with low arousal, the lateral but not the mediobasal hypothalamus is overactivated, the anti-aggressive effects of serotonergic neurotransmission appear lost, and vasopressinergic neurotransmission decreases.

[3] Unfortunately, a similar table cannot be compiled for the lateral hypothalamus because of the paucity of data.

Table 4.2 Functional changes in mediobasal hypothalamic regions in models of abnormal aggression and aggression-related psychopathologies

Model	c-Fos	5-HT	AVP	NK1	GLU	GABA	DA	Arousal
Postweaning social isolation	⇑	⇓*						High
Selection for low anxiety	⇑							High
Early maternal separation		⇓	⇔					High
Cocaine in adolescence	⇑		⇑			⇑		High
Amphetamine in adolescence			⇑					High
Psychopaths (moderate level)	⇑†*							High††
Borderline personality disorder	⇑‡	⇓*						High‡‡
Selection for high anxiety	⇔							Low
Selection for high aggression	⇔					⇓		Low
Glucocorticoid deficiency	⇔			⇑				Low
Social subjugation in adolescence		⇑*	⇓					Low
Anabolic steroid during adolescence	⇑	⇓	⇑		⇑	⇑	⇑	Unknown
Acute alcohol		⇑*						Unknown

Models grouped according to emotional background
⇑, increase; ⇔, no change; ⇓, decrease; *empty cells*, no data available; *, whole hypothalamus; †, activation evaluated by fMRI; ††, increased anger during punishment/retaliation testing; ‡, increased gray matter volume; ‡‡, general feature (not investigated in the study); *high arousal*, high basal glucocorticoid levels or high glucocorticoid response to aggression and/or high basal heart rates or high autonomic response to aggression; *low arousal*, the opposite pattern; *unknown*, not investigated in the specific conditions of the model

Thus, the hypothalamic backgrounds of various forms of abnormal aggression show qualitative differences, suggesting that there are alternative "neurobiological roads" to abnormal aggression.

4.2 The Prefrontal Cortex

4.2.1 Prefrontal Anatomy of Violence: A Prelude

The prefrontal cortex is the headquarters of higher judgment that hosts the neuronal fingerprints of personality and temperament (DeYoung et al. 2010; Whittle et al. 2006). Its role in shaping aggressive personalities was recognized rather early mainly based on personality changes that follow its partial destruction; such studies were pioneered by the famous nineteenth-century case of Phineas Gage (Blumer and Benson 1975). Brain imaging studies enhanced understanding by establishing associations between structural and/or functional deficits located within the prefrontal cortex on one side and aggression histories and aggression-related psychopathologies on the other. The two types of studies nicely complement each other; while prefrontal lesions increase aggression, brain imaging studies reveal prefrontal deficits in aggressive people. The coherence and significance of these findings propelled the area in the forefront of aggression research, and the inhibitory control of aggression by the prefrontal cortex became a cornerstone of understanding aggression control. In sharp contrast with these human findings, aggressive behavior activates the prefrontal cortex in animals; moreover, the activation is stronger in models of abnormal aggression. These findings cannot be ascribed to species differences because very similar findings were obtained in humans. While a *history* of aggression does correlate with prefrontal deficits, aggressive behaviors performed *during* brain imaging activate the prefrontal cortex; moreover, activations correlate positively with the level of aggressiveness. In addition to this marked discrepancy, a closer look shows that brain lesion and brain imaging studies are not as consistent as generally thought. The location of aggression-inducing prefrontal damages—accounted for by clinical investigations—and aggression-related prefrontal deficits, as revealed by brain imaging studies, does not overlap as a rule. The same is valid for brain imaging studies themselves: very different patterns of prefrontal dysfunctions were seen in highly similar subjects studied by highly similar techniques and sometimes by overlapping sets of investigators. While deficits were almost always seen within the large brain area called prefrontal cortex, every study revealed a unique patchwork of deficient regions. Overall, one cannot question the role of the prefrontal cortex in aggression control, and this brain area may indeed bear the fingerprints of aggressive personalities. However, the fingerprint is rather blurred at present.

Due to such discrepancies, this section will have a specific structure. First, we will overview the general view, after which we will cross-check them with the available findings. This will be followed by an attempt to create an alternative view

4.2.2 An Overall Summary of the General View

Arguments supporting the view that the prefrontal cortex inhibits aggression are apparently overwhelming. The main lines of reasoning are as follows:

1. Lesions placed within this brain area increase aggression in animals (*rats*: de Bruin et al. 1983; Kolb 1974; Rudebeck et al. 2007; *monkeys*: Izquierdo 2005; Machado and Bachevalier 2006), and similarly, accidental prefrontal lesions result in increased aggression in humans (Anderson et al. 1999; Barrash et al. 2000, 2011; Bear and Fulop 1987; Blair and Cipolotti 2000; Blumer and Benson 1975; Luria 1980; Max et al. 2006; Pardini et al. 2011; Sarazin et al. 2003; Shamay-Tsoory et al. 2010; Tranel et al. 2002). In addition, prefrontal damage leads in humans to increased anger and irritability, impulsivity, deficits in recognizing the intentions of social partners, as well as indifference to rules and lack of moral reasoning, all being features that increase the likelihood of engagement in aggressive acts (Anderson et al. 1999; Bear and Fulop 1987; Berlin et al. 2004, 2005; Blumer and Benson 1975; Sarazin et al. 2003; Shamay-Tsoory et al. 2010).

2. Brain imaging studies show prefrontal deficits in aggressive people. Such deficits were seen in (1) aggression-related psychopathologies (e.g., antisocial and borderline personality disorders, disruptive behavior, psychopathy) (Birbaumer et al. 2005; Boccardi et al. 2011; Carrasco et al. 2012; Ermer et al. 2013; Fahim et al. 2011; Goethals et al. 2005; Goyer et al. 1994; Gregory et al. 2012; Hazlett et al. 2005; Hirono et al. 2000; New et al. 2002; Raine et al. 1992, 1994, 1997, 2000, 2011; Soloff et al. 2003; Tiihonen et al. 2008), (2) violent patients showing other disorders (*drug abuse*: Payer et al. 2011; *epilepsy*: Juhász et al. 2001; *mental retardation*: Critchley et al. (2000); *various psychiatric disorders; patients studied together*: Antonucci et al. 2006; Gansler et al. 2011; Volkow et al. 1995), and (3) violent offenders (Amen et al. 1996; Boes et al. 2008; Ducharme et al. 2011; Mehta and Beer 2010; Meyer-Lindenberg et al. 2006; Pardini and Phillips 2010; Passamonti et al. 2006). Findings in animals are in conflict with human data in this respect, but basic strategies in the two research areas are substantially different. This issue will be discussed below.
3. Aggressiveness and aggression-related psychopathologies were associated with poor performance in psychiatric tests of frontal lobe functions (Bergvall et al. 2001; Best et al. 2002; Deckel et al. 1996; Dinn and Harris 2000; Dolan and Park 2002; Dowson et al. 2004; Lapierre et al. 1995).
4. Prefrontal deficits were associated with psychological features, which likely increase the propensity to behave aggressively: poor ability to recognize emotions or the valence of social interactions, poor cognitive control over behavior, inability to take appropriate moral decisions, and feelings like anger, social anxiety, and impulsiveness (Bergvall et al. 2001; Best et al. 2002; Birbaumer et al. 2005; Deckel et al. 1996; Dowson et al. 2004; Dinn and Harris 2000; Fulwiler et al. 2012; Gansler et al. 2011; Goyer et al. 1994; Heekeren et al. 2003; Matsuo et al. 2009; Pardini and Phillips 2010; Passamonti et al. 2012; Potts et al. 2006; Sala et al. 2011; Sheng et al. 2010; Spitzer et al. 2007; Soloff et al. 2003). Impulsivity, social anxiety, cognitive set shifting, inhibitory control, and decision making were all associated with prefrontal functions in animals; thus, human and animal studies provide quite consistent findings in this respect (Cambon et al. 2010; Chudasama et al. 2012; Dalley et al. 2002; Kamigaki et al. 2012; Paine et al. 2013; Prinssen et al. 2012; Sagvolden 2006).

The findings briefly reviewed above strongly point to the conclusion that the prefrontal cortex exerts an inhibitory control over aggression. Overall, the conclusions of animal and human studies are quite similar and together resulted in the "prefrontal deficit theory" of aggression.

4.2.3 *Problems with the General View*

4.2.3.1 Size

The findings summarized above seem harmonious at the first sight, but a closer scrutiny reveals that they are fundamentally incongruent. Problems start already with the size of this brain area. While the human prefrontal cortex is delimited in various ways by various authors (for contrasting delimitations, see Critchley et al. 2000; Barrash et al. 2000; Raine et al. 1997, 2011; Davidson et al. 2000), it can be considered very large according to any of the delimitations. For the purposes of this review, we summed up the various views of the authors reviewed here and corroborated these with the brain atlas of Mai et al. (1998) (Fig. 4.5a). As seen on the figure, the brain area generally considered the "prefrontal cortex" makes up more than one-third of the neocortex. It is rather difficult to imagine how such a large brain area can as a whole project to particular subcortical structures and influence their functions as a whole.

The prefrontal cortex of rats is relatively smaller. The widespread consensus is that it is composed of the medial and orbitofrontal prefrontal cortices (Fig. 4.5b). Some authors also include the agranular insular and piriform cortices. Even if smaller relatively, it still comprises approximately one-fifth of the rat neocortex. Thus, it is still too large to assume that it is functionally homogeneous and affects subcortical structure as a whole.

These considerations show that the prefrontal cortex is not a unit that can be considered as such when mechanisms underlying aggressiveness are evaluated. As a phrase, it is just as general as "subcortical structures" or "limbic system." Aggression cannot be controlled by the whole prefrontal cortex; projections that can be physically accommodated by lower structures must come from restricted regions of this vast area. The question arises whether the available data allow the delimitation of aggression-specific regions in the prefrontal cortex. A cross-checking of findings obtained with various technologies shows that they cannot.

4.2.3.2 Brain Lesion and Brain Imaging Studies Do Not Support Each Other

One of the strong arguments supporting the "prefrontal deficit theory" of aggression is that lesions of the prefrontal cortex induce a state called "acquired psychopathy," while psychopaths show structural and functional deficits exactly in the prefrontal cortex. However, lesions leading to "acquired psychopathy" are located elsewhere than the structural and functional alterations seen in aggression-related psychopathologies.

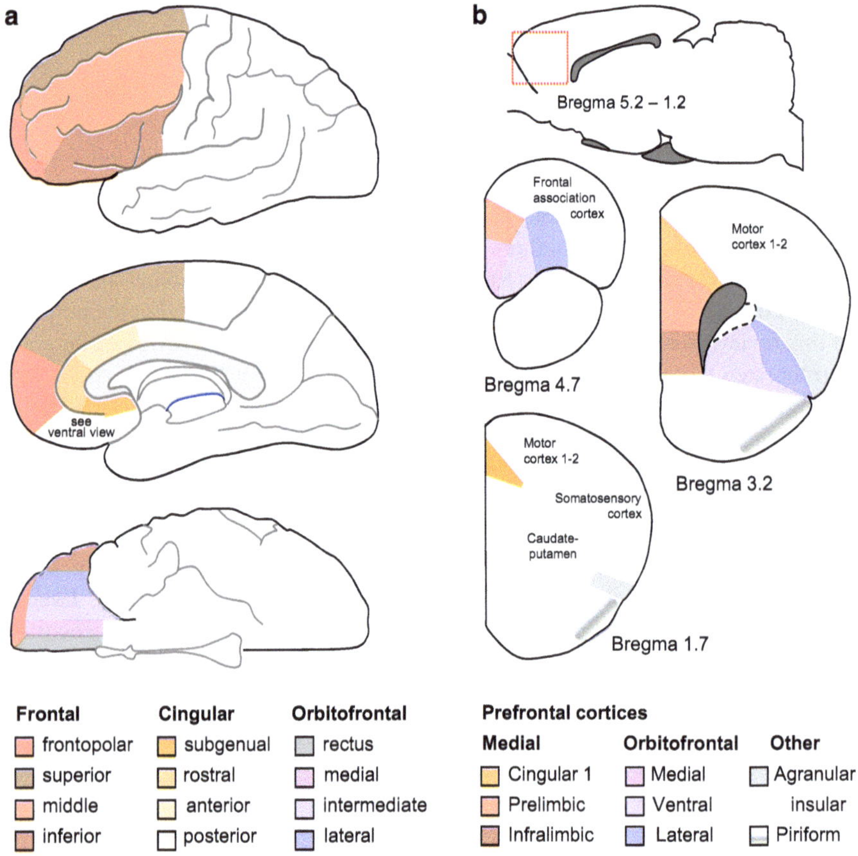

Fig. 4.5 *The prefrontal cortex and its main parts in humans* (**a**) *and rats* (**b**). The *red dotted square* on the upper schematic of panel **b** shows the approximate location in the rat brain. Only parts of this region are considered "prefrontal" (see sections through the rat brain)

Findings in Humans

Prefrontal lesions elicit the development of psychopathic-like traits only if they include the orbitofrontal area (Anderson et al. 1999; Barrash et al. 2000, 2011; Bear and Fulop 1987; Berlin et al. 2004, 2005; Blair and Cipolotti 2000; Blumer and Benson 1975; Pardini et al. 2011; Sarazin et al. 2003; Shamay-Tsoory et al. 2010; Tranel et al. 2002). Lesions located elsewhere in the prefrontal cortex either did not induce personality changes or induced weak changes (Barrash et al. 2000; Berlin et al. 2005; Pardini et al. 2011; Sarazin et al. 2003; Shamay-Tsoory et al. 2010). Moreover, there are a few studies suggesting that damages located in the dorsolateral or dorsomedial regions are associated with quite different behavioral changes, e.g., motor dysfunctions and apathy (Cummings 1995; Luria 1980; Niedermeyer 1998; Sarazin et al. 2003; Tekin and Cummings 2002). We found only one study where dorsolateral prefrontal lesions were paralleled by psychopathic-like personality changes (Max et al. 2006). However, patients had multiple brain lesions in this study,[4] which makes the specificity of dorsolateral prefrontal regions questionable.

Taken together, the findings reviewed above show that psychopathic-like traits are associated with orbitofrontal lesions. In sharp contrast, there are quite many studies where orbitofrontal deficits were not seen in people suffering from aggression-related psychopathologies including psychopathy (Antonucci et al. 2006; Boes et al. 2008; Ducharme et al. 2011; Fahim et al. 2011; Goethals et al. 2005; Gregory et al. 2012; Hirono et al. 2000; Juengling et al. 2003; Juhász et al. 2001; Laakso et al. 2002; Rosell et al. 2010). In these studies, deficits were seen in other prefrontal regions, e.g., in the dorsolateral prefrontal cortex.

Findings in Animals

In animals, aggression increased when the orbitofrontal cortex but not when other prefrontal areas e.g., the medial prefrontal cortex or the anterior cingulate were damaged (de Bruin et al. 1983; Izquierdo 2005; Kolb 1974; Machado and Bachevalier 2006; Rudebeck et al. 2007). While aggression was increased specifically by damages located in the orbitofrontal cortex, changes associated with abnormal forms of aggression seem to be located elsewhere. The issue is certainly poorly studied, but it was shown that postweaning social isolation reduced the volume of the medial prefrontal cortex (Schubert et al. 2009; Day-Wilson et al. 2006). Although data are sparse, they tentatively suggest that the dissociation between the location of aggression-enhancing brain lesions and aggression-related prefrontal changes is different in rodents as well.

4.2.3.3 Brain Imaging Studies Do Not Support Each Other

These studies do not support each other in the meaning that the deficits observed by different authors are located in anatomically distinct regions of the prefrontal cortex.

[4] Lesions were observed in the basal ganglia, corpus callosum, occipital lobe, operculum, orbitofrontal gyrus, parietal lobe, rectal gyrus (part of the orbitofrontal area), temporal lobe, thalamus, as well as prefrontal white matter that likely included the projections of the orbitofrontal area.

Human Studies

The available studies do show aggression-related deficits in the vast area called "prefrontal cortex," but the precise location and size of deficient areas varies from study to study. As examples, we summarized the findings of Boccardi et al. (2011) and Tiihonen et al. (2008) in Fig. 4.6a. The subjects of both studies were violent offenders, were diagnosed with antisocial personality disorder, and showed psychopathic traits. In addition, the populations studied included murderers in rather similar proportions and controls were "healthy" (nonaggressive) individuals. Both papers identified gray matter deficits in prefrontal regions. A closer look, however, shows that gray matter deficits were seen in rather different areas; overlaps are meager. The lack of similarities in affected brain areas is even more blatant in two studies on gray matter deficits associated with psychopathic traits (Ermer et al. 2012; Gregory et al. 2012). Subjects were violent offenders (including murderers), were diagnosed with antisocial personality disorder, and showed psychopathic traits in both cases. The approaches were somewhat different: Ermer et al. (2013) studied the correlation between psychopathic traits and gray matter volumes, while Gregory et al. (2012) compared the gray matter of antisocial personality-disordered violent offenders who did or did not show psychopathic traits. However, the ultimate aim of the two studies was similar: both sets of authors aimed at identifying gray matter deficits associated with psychopathic traits. Both studies found prefrontal deficits, but in completely different areas of the prefrontal cortex (Fig. 4.6b). In addition to the total lack of overlaps, Ermer et al. (2013) found a gray matter "gain" in the frontopolar area that was not seen by Gregory et al. (2012). Data on prefrontal deficits associated with borderline personality disorder are also divergent. Figure 4.6c compares the data obtained by Goethals et al. (2005), Soloff et al. (2003), and Juengling et al. (2003). All three evaluated local brain metabolic differences at resting state; all three found deficits compared to healthy controls, but again, the location of the deficits was rather different.

Intriguing discrepancies can be observed even in studies where the regions of interest (the areas studied as a whole) were very large and where overlaps were theoretically unavoidable. Here we compared three articles published by overlapping sets of authors (Raine et al. 2000, 2011; Yang et al. 2005). The first paper showed that the volume of the prefrontal cortex—defined as all gray matter anterior to the genu callosum—is reduced in violent offenders (including murderers) who were diagnosed with antisocial personality disorder and showed psychopathic traits (Fig. 4.7a). Differences were smaller in the last study of the series (Raine et al. 2011). Even more intriguingly, Yang et al. (2005) found no prefrontal gray or white mater alterations in antisocial personality-disordered subjects showing psychopathic traits (Fig. 4.7b). By contrast, differences were obtained when pathological liars were compared with non-liars. As all the subjects of this triad of studies were antisocial personality disordered and psychopathic, the findings of Raine et al. (2000) and Yang et al. (2005) combined suggest that violent offending (including murder) alone and lying alone are associated with similar prefrontal deficits, while antisocial personality combined with psychopathic traits is not.

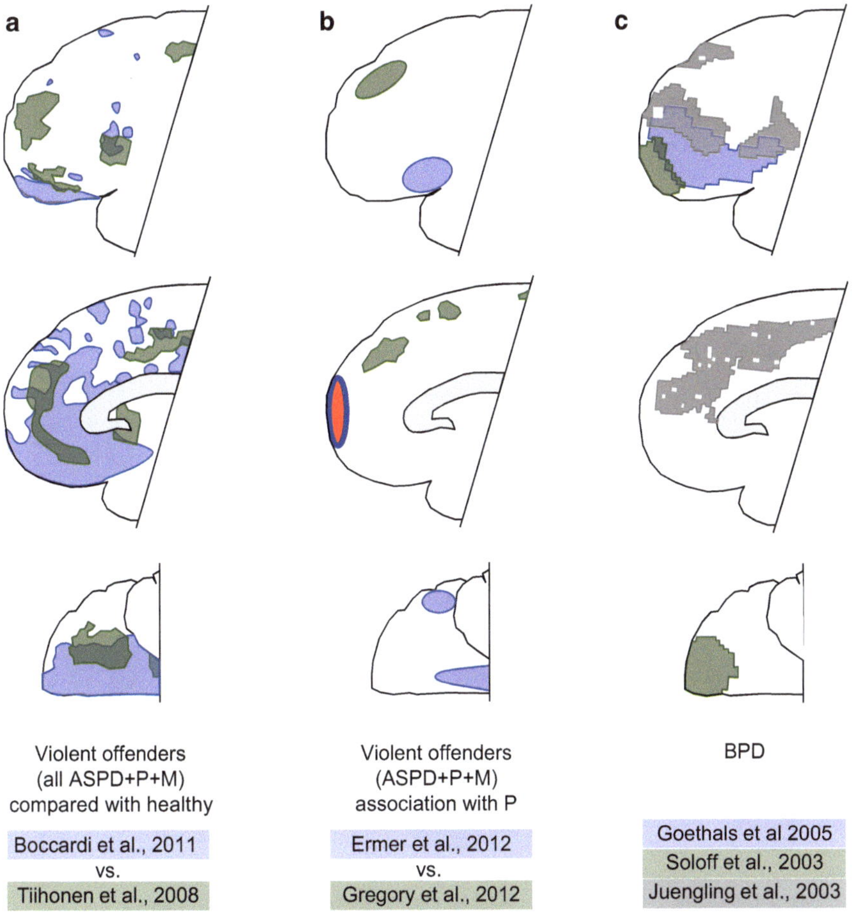

Fig. 4.6 *The location of brain deficits in very similar studies*. (**a**) violent offenders; (**b**) psychopathic traits; (**c**) borderline personality disorder. The characteristics of patients and the authors of studies are shown below panels. *Abbreviations*: *ASPD*, antisocial personality disorder; *P*, psychopathic traits; *M*, murderers included in the study; *BPD*, borderline personality disorder; *red area*, gray matter increase observed by Ermer et al. (2012); *other colored areas*, deficient regions; color codes identify the authors of the studies compared within panels

Besides these blatant examples, we compared the location and size of various prefrontal deficits in all the studies where these were clearly identifiable and where the conditions of the studies were comparable. Findings were similar in three studies only, all three involving aggressive children. The subjects of two of these studies were healthy children with conduct problems (Boes et al. 2008; Ducharme et al. 2011) (Fig. 4.7c). The third study was on conduct-disordered children (Fahim et al. 2011). Deficits were seen in highly overlapping regions of the cingulate cortex; neither study found orbitofrontal deficits (Fig. 4.7c). At this stage, this seems the only replicable finding of brain imaging studies.

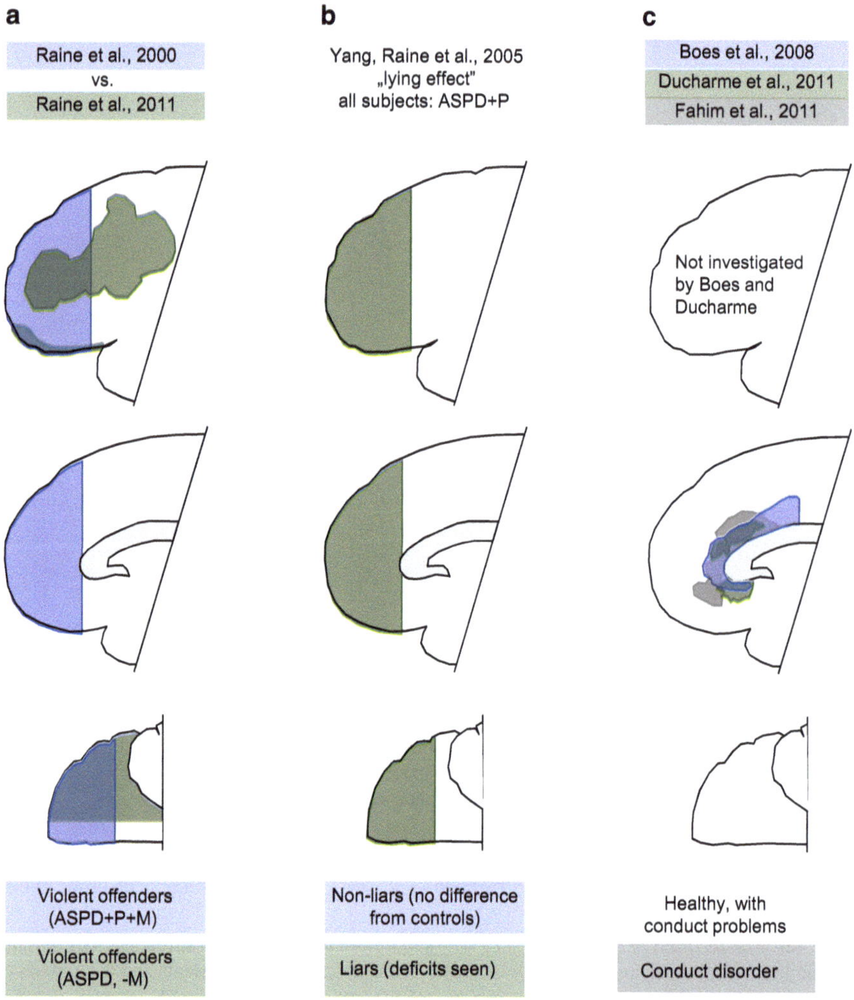

Fig. 4.7 *Prefrontal deficits in antisocial personality disordered (ASPD) adults and children showing conduct problems.* (**a**), (**b**) ASPD patients with different characteristics. *+P*, psychopathic traits; *+M*, murderers included in the study; *–M*, murderers not included; *lack of blue color on panel **b***, lack of prefrontal dysfunctions in ASPD+P subjects. (**c**) Prefrontal deficits in healthy children with conduct problems (Boes et al. 2008; Ducharme et al. 2011) and in conduct disordered children (Fahim et al. 2011)

One can argue that the sophisticated nature of brain imaging techniques and the subtleties of statistics that need to be employed preclude the precise overlapping of the regions where deficits are seen. However, the cross-checking of findings at a lower resolution strengthens the view that while prefrontal deficits are commonly seen, the location of these is highly variable (see Table 4.3. for findings grouped according to Brodmann areas).

Table 4.3 Prefrontal changes observed by various authors, grouped according to Brodmann areas

Area	Decreased	Increased	No change	Disorder	Technique	References
BA 4	X			ASPD	sMRI	Raine et al. (2011)
	X			ASPD + PP	sMRI	Tiihonen et al. (2008)
	X			PP	sMRI	Ermer et al. (2013)
			X	ASPD − PP	sMRI	Gregory et al. (2012)
			X	ASPD + PP	sMRI	Gregory et al. (2012)
BA 8	X			ASPD + PP	sMRI	Tiihonen et al. (2008)
	X			DBD	sMRI	Fahim et al. (2011)
		X		BPD	PET	Soloff et al. (2008)
			X	ASPD	sMRI	Raine et al. (2011)
			X	ASPD − PP	sMRI	Gregory et al. (2012)
			X	BPD	sMRI	Hazlett et al. (2005)
			X	BPD	PET	Soloff et al. (2005)
			X	BPD	sMRI	Tebartz van Elst et al. (2003)
			X	BPD-IED	PET	New et al. (2009)
			X	PP	sMRI	Ermer et al. (2013)
BA 9	X			ASPD	sMRI	Raine et al. (2000)
	X			ASPD	sMRI	Raine et al. (2011)
	X			ASPD − PP	sMRI	Gregory et al. (2012)
	X			BPD	PET	Soloff et al. (2003)
			X	ASPD	sMRI	Yang et al. (2005)
			X	ASPD − PP	sMRI	Gregory et al. (2012)
			X	ASPD + PP	sMRI	Tiihonen et al. (2008)
			X	BPD	sMRI	Hazlett et al. (2005)
			X	BPD	PET	Soloff et al. (2005)
			X	BPD	sMRI	Tebartz van Elst et al. (2003)
			X	BPD-IED	PET	New et al. (2009)
			X	PP	sMRI	Ermer et al. (2013)
BA 10	X			ASPD	sMRI	Raine et al. (2011)
	X			ASPD	sMRI	Raine et al. (2000)
	X			ASPD/BPD	SPECT	Goethals et al. (2005)
	X			ASPD + PP	sMRI	Gregory et al. (2012)
	X			ASPD + PP	sMRI	Tiihonen et al. (2008)
	X			BPD	PET	Soloff et al. (2003)
	X			BPD	PET	Soloff et al. (2003)
	X			PP	sMRI	Boccardi et al. (2011)
		X		PD	PET	New et al. (2002)
		X		PP	sMRI	Ermer et al. (2013)
			X	ASPD	sMRI	Yang et al. (2005)
			X	ASPD − PP	sMRI	Gregory et al. (2012)
			X	ASPD + PP	sMRI	Gregory et al. (2012)
			X	BPD	sMRI	Hazlett et al. (2005)
			X	BPD	sMRI	Tebartz van Elst et al. (2003)

(continued)

Table 4.3 (continued)

Area	Decreased	Increased	No change	Disorder	Technique	References
			X	BPD-IED	PET	New et al. (2009)
			X	PD	PET	New et al. (2002)
BA 11	X			ASPD	sMRI	Raine et al. (2000)
	X			ASPD	sMRI	Raine et al. (2011)
	X			ASPD + PP	sMRI	Tiihonen et al. (2008)
	X			BPD	PET	Soloff et al. (2003)
	X			BPD	PET	Soloff et al. (2005)
	X			BPD	sMRI	Tebartz van Elst et al. (2003)
	X			DBD	sMRI	Fahim et al. (2011)
	X			PP	sMRI	Boccardi et al. (2011)
	X			PP	sMRI	Ermer et al. (2013)
			X	Healthy	sMRI	Boes et al. (2008)
			X	Healthy	sMRI	Ducharme et al. (2011)
			X	ASPD	sMRI	Yang et al. (2005)
			X	ASPD − PP	sMRI	Gregory et al. (2012)
			X	ASPD + PP	sMRI	Gregory et al. (2012)
			X	BPD	sMRI	Morandotti et al. (2013)
			X	BPD	sMRI	Hazlett et al. (2005)
			X	BPD-IED	PET	New et al. (2009)
			X	PD	PET	New et al. (2002)
BA 12	X			ASPD	sMRI	Raine et al. (2011)
	X			ASPD	sMRI	Raine et al. (2000)
	X			ASPD + PP	sMRI	Tiihonen et al. (2008)
	X			BPD	sMRI	Tebartz van Elst et al. (2003)
	X			DBD	sMRI	Fahim et al. (2011)
			X	Healthy	sMRI	Boes et al. (2008)
			X	Healthy	sMRI	Ducharme et al. (2011)
			X	ASPD	sMRI	Yang et al. (2005)
			X	ASPD − PP	sMRI	Gregory et al. (2012)
			X	ASPD + PP	sMRI	Gregory et al. (2012)
			X	BPD	sMRI	Hazlett et al. (2005)
			X	BPD-IED	PET	New et al. (2009)
BA 24	X			Healthy	sMRI	Boes et al. (2008)
	X			Healthy	sMRI	Ducharme et al. (2011)
	X			ASPD	sMRI	Raine et al. (2000)
	X			BPD	sMRI	Hazlett et al. (2005)
	X			BPD	sMRI	Tebartz van Elst et al. (2003)
	X			DBD	sMRI	Fahim et al. (2011)
	X			PP	sMRI	Boccardi et al. (2011)
		X		PD	PET	New et al. (2002)
			X	ASPD	sMRI	Yang et al. (2005)
			X	ASPD − PP	sMRI	Gregory et al. (2012)
			X	ASPD + PP	sMRI	Gregory et al. (2012)

(continued)

Table 4.3 (continued)

Area	Decreased	Increased	No change	Disorder	Technique	References
			X	ASPD + PP	sMRI	Tiihonen et al. (2008)
			X	BPD-IED	PET	New et al. (2009)
			X	PD	PET	New et al. (2002)
			X	PP	sMRI	Ermer et al. (2013)
BA 25	X			Healthy	sMRI	Boes et al. (2008)
	X			Healthy	sMRI	Ducharme et al. (2011)
	X			PP	sMRI	Boccardi et al. (2011)
			X	ASPD − PP	sMRI	Gregory et al. (2012)
			X	ASPD + PP	sMRI	Gregory et al. (2012)
			X	ASPD + PP	sMRI	Tiihonen et al. (2008)
			X	BPD-IED	PET	New et al. (2009)
			X	PD	PET	New et al. (2002)
			X	PP	sMRI	Ermer et al. (2013)
BA 32	X			ASPD	sMRI	Raine et al. (2000)
	X			ASPD + PP	sMRI	Tiihonen et al. (2008)
	X			BPD	PET	Soloff et al. (2008)
	X			PP	sMRI	Boccardi et al. (2011)
			X	Healthy	sMRI	Boes et al. (2008)
			X	Healthy	sMRI	Ducharme et al. (2011)
			X	ASPD	sMRI	Yang et al. (2005)
			X	ASPD − PP	sMRI	Gregory et al. (2012)
			X	ASPD + PP	sMRI	Gregory et al. (2012)
			X	BPD	sMRI	Hazlett et al. (2005)
			X	BPD	sMRI	Tebartz van Elst et al. (2003)
			X	BPD-IED	PET	New et al. (2009)
			X	PD	PET	New et al. (2002)
			X	PP	sMRI	Ermer et al. (2013)
BA 44	X			ASPD + PP	sMRI	Tiihonen et al. (2008)
	X			PP	sMRI	Boccardi et al. (2011)
			X	ASPD	sMRI	Raine et al. (2011)
			X	ASPD − PP	sMRI	Gregory et al. (2012)
			X	ASPD + PP	sMRI	Gregory et al. (2012)
			X	BPD	sMRI	Morandotti et al. (2013)
			X	BPD-IED	PET	New et al. (2009)
			X	PP	sMRI	Ermer et al. (2013)
BA 45	X			ASPD + PP	sMRI	Tiihonen et al. (2008)
			X	ASPD − PP	sMRI	Gregory et al. (2012)
			X	ASPD + PP	sMRI	Gregory et al. (2012)
			X	BPD	sMRI	Morandotti et al. (2013)
			X	BPD	sMRI	Tebartz van Elst et al. (2003)
			X	BPD-IED	PET	New et al. (2009)
			X	PD	PET	New et al. (2002)
			X	PP	sMRI	Ermer et al. (2013)

(continued)

Table 4.3 (continued)

Area	Decreased	Increased	No change	Disorder	Technique	References
BA 46	X			ASPD	sMRI	Raine et al. (2011)
	X			ASPD	sMRI	Raine et al. (2000)
	X			PP	sMRI	Boccardi et al. (2011)
		X		PD	PET	New et al. (2002)
			X	ASPD	sMRI	Yang et al. (2005)
			X	ASPD	sMRI	Raine et al. (2011)
			X	ASPD + PP	sMRI	Tiihonen et al. (2008)
			X	ASPD + PP	sMRI	Gregory et al. (2012)
			X	ASPD − PP	sMRI	Gregory et al. (2012)
			X	BPD	sMRI	Tebartz van Elst et al. (2003)
			X	BPD	sMRI	Hazlett et al. (2005)
			X	BPD-IED	PET	New et al. (2009)
			X	PD	PET	New et al. (2002)
			X	PP	sMRI	Ermer et al. (2013)
BA 47	X			BPD	sMRI	Tebartz van Elst et al. (2003)
	X			BPD	sMRI	Morandotti et al. (2013)
	X			DBD	sMRI	Fahim et al. (2011)
	X			PP	sMRI	Boccardi et al. (2011)
	X			PP	sMRI	Ermer et al. (2013)
			X	Healthy	sMRI	Ducharme et al. (2011)
			X	ASPD − PP	sMRI	Gregory et al. (2012)
			X	ASPD + PP	sMRI	Tiihonen et al. (2008)
			X	ASPD + PP	sMRI	Gregory et al. (2012)
			X	BPD	sMRI	Hazlett et al. (2005)
			X	BPD-IED	PET	New et al. (2009)
			X	PD	PET	New et al. (2002)

The referenced authors employed various techniques; the phrases "decreased," "increased," and "no change" should be understood in terms typical to the technique. For example, "decrease" means decreased volume in structural MRI studies, low glucose metabolism in PET studies, etc

Abbreviations: *ASPD*, antisocial personality disorder; *BPD*, borderline personality disorder; *DBD*, disruptive behavior disorder; *healthy*, healthy subjects with a predilection towards aggressiveness; *PD*, various personality disorders (not differentiated); *PP*, psychopathy; +, present; −, absent. Note that subjects were aggressive in all these studies

Animal Studies

Structural and functional brain imaging studies are scarce in the animal literature, and the available ones cannot be used to compare the location and extent of prefrontal deficits that are associated with aggression.

4.2.3.4 "Aggression History" and "Manifest aggression" Studies Do Not Support Each Other

All the studies reviewed above belong to the "aggression history" type where brain imaging is performed while subjects lie still in the apparatus; aggressiveness is attributed to them based on their past actions. In the majority of such subjects, prefrontal deficits are seen, even if the location of these is highly variable. In manifest aggression studies, where subjects perform aggression-related tasks during brain imaging, prefrontal *activations* are seen; moreover, the magnitude of activation correlates positively with the history of violence. The two types of studies do not support each other in the meaning that while deficits are seen in the first type, activations are seen in the second.

Findings in Humans

Specific subareas of the prefrontal cortex were activated by the following: (1) stimuli that potentially elicit aggression, e.g., the detection of fear, anger, and threatening situations (Blair et al. 1999; Dougherty et al. 1999; Hermans et al. 2008; Pardini and Phillips 2010; Sugiura et al. 2009); (2) attitudes and feelings that increase the likelihood of aggression, e.g., negative affect, psychological distress, frustration, and anger (Abler et al. 2005; Herpertz et al. 2001; Kimbrell et al. 1999; Li et al. 2006; Schneider et al. 2000; Siegrist et al. 2005); and (3) imagined aggression, playing aggressive video games and delivering punishment in competitive situations (aggressive retaliation) (King et al. 2006; Lotze et al. 2007; Montag et al. 2012; New et al. 2009; Strenziok et al. 2011; Veit et al. 2010).

Thus, prefrontal activations paralleled all three main stages of aggressive conflicts: the detection of threat, the induction of feelings that engender aggressiveness, and the execution of aggressive acts. Intriguingly, prefrontal activations were enhanced in patients suffering from aggression-related psychopathologies, e.g., personality disorders, intermittent explosive disorder, and psychopathy (Herpertz et al. 2001; New et al. 2009; Schneider et al. 2000; Völlm et al. 2010). In addition, stronger activation was seen in healthy subjects who were especially aggressive in the experimental task (e.g., excessive first-person-shooter video players, cocaine abusers, and chronically violent men) (Li et al. 2006; Montag et al. 2012; Pardini and Phillips 2010). Strikingly, the orbitofrontal area which increases aggression when damaged was quite often activated in these studies (Blair et al. 1999; Herpertz et al. 2001; Kimbrell et al. 1999; King et al. 2006; New et al. 2009; Strenziok et al. 2011; Sugiura et al. 2009; Veit et al. 2010).

The anatomical substrates of such activations are just as variable as those seen in aggression history studies. For example, eight studies reported activations within the orbitofrontal area (see above); the anterior cingulate was activated in six studies, out of which only one reported the concomitant activation of the orbitofrontal cortex (Abler et al. 2005; Blair et al. 1999; Dougherty et al. 1999; Li et al. 2006;

Siegrist et al. 2005). In contrast with "aggression history" studies, however, manifest aggression studies are rather few in number, and it is difficult to find two that would allow a point-by-point comparison. Even when the basic tool to study aggression was similar—e.g., aggressive video games—the circumstances were quite different. In one publication the processing of screenshot images was studied, while the other focused on playing per se (King et al. 2006; Montag et al. 2012). One can assume that differences in the location of activated areas are explained by such differences in methodologies and circumstances. Naturally, the issue needs further studies, but at present stage, a "prefrontal activation theory" of aggression seems to take shape, especially because this theory is strongly supported by animal findings.

Findings in Animals

Although orbitofrontal lesions decrease aggression in rodents just like in humans, aggressive behaviors do activate the prefrontal cortex; moreover, activation was enhanced in models where subjects showed abnormal forms of aggression.

Resident-intruder tests performed in normal subjects either activated all the prefrontal areas that were investigated (*rats*: Halász et al. 2006; *mice*: Haller et al. 2006; Wall et al. 2012) or activated extended subregions of these areas (*rats*: Beiderbeck et al. 2012 Clinton et al. 2011; *mice*: Duncan et al. 2009). The medial prefrontal cortex was also activated by maternal aggression in rats (Caffrey et al. 2010) and play fighting in hamsters (Cheng et al. 2008). These studies quantified the expression of immediate early genes, for which they may be criticized for the low temporal resolution of this technology. Indeed, the findings do not exclude the possibility that the prefrontal cortex was activated by the inhibition rather than by the initiation of aggressive acts. Biting attacks occur in bursts in rodents, which are separated by episodes of relatively peaceful interactions. The development of the c-Fos signal takes time, for which any of the events of the aggressive conflict can cause the activations seen post hoc. Therefore, the effects of escalations and returns to peacefulness cannot be told apart. A few studies show, however, that the activation of the prefrontal cortex is not related to temporary limitations of aggression. For example, Ferris et al. (2008) employed brain imaging techniques in rats that were visually challenged by opponents. Although animals did not perform aggressive behavior (i.e., there was no aggression to be "stopped" by the prefrontal cortex), all the subregions of the prefrontal cortex were activated. Similar findings were obtained by Halász et al. (2006) who showed that the prelimbic, anterior cingular, agranular insular, and ventral orbitofrontal cortices were activated by psychosocial encounters, i.e., by a situation where intruders and residents were physically separated (i.e., fights were prevented). Psychosocial encounters activated the prefrontal cortex in mice as well (Ago et al. 2013). Taken together, these findings preclude that increased prefrontal activations are specifically related to temporary (within-encounter) dampening of aggressive behavior.

Even more strikingly, prefrontal activation was exacerbated in models where the subjects showed abnormal forms of aggression. In rats selected for anxiety, for instance, the prelimbic, infralimbic, and anterior cingulate cortices were all activated in addition to the ventral orbitofrontal cortex that was activated by fights already in controls (i.e., in rats not selected for anxiety) (Beiderbeck et al. 2012). In the study by Halász et al. (2006), glucocorticoid dysfunction-induced abnormal aggression correlated negatively with the activation of prefrontal GABAergic, but positively with the activation of prefrontal glutamatergic neurons, the latter being the neuron type that sends projections to subcortical structures. This study not only suggested that the outputs of the prefrontal cortex were upregulated in abnormally aggressive subjects but also showed that local inhibitory neurons were downregulated. Postweaning social isolation, a model of exacerbated emotional aggression increased the activation of the prefrontal cortex especially in the medial and lateral orbitofrontal cortices and the anterior cingulate cortex (*mice*: Ago et al. 2013; *rats*: Toth et al. 2012). Finally, mice genetically selected for aggressive behavior showed larger activations in both the medial and the orbitofrontal prefrontal cortices as compared to their more peaceful counterparts (Haller et al. 2006). There are only two models of abnormal aggression where similar phenomena were not observed and where prefrontal activation was decreased by fights as compared to normally fighting controls. In the pubertal stress model, the aggression-induced activation of the medial orbitofrontal cortex was blunted (Márquez et al. 2013). No differences were seen in other prefrontal areas. In the study by Wall et al. (2012), rats were socially isolated for 4 weeks after weaning, and the aggression test was performed in late adolescence (postnatal days 49–52). The activation of all three medial prefrontal regions was blunted in socially isolated as compared to socially housed rats. Even in these cases, however, the prefrontal cortex *was* activated by fights.

4.2.3.5 Summary

The findings reviewed above can be summarized as follows: (1) prefrontal damages increase aggression only when they include orbitofrontal areas. The very same region is not *necessarily* dysfunctional in aggressive individuals. (2) Non-orbitofrontal damages do not increase aggression. Aggressiveness was *often* specifically associated with non-orbitofrontal dysfunctions. (3) Aggression-related prefrontal dysfunctions occur in patches. The location and extent of dysfunctional "patches" vary from study to study. Findings are discordant even when the characteristics of subjects and the methods of study are highly similar. (4) The execution of aggressive acts activates the prefrontal cortex. Activation is stronger in abnormally aggressive individuals. (5) Human and animal findings are highly consistent. Despite their rather simplistic formulation, the statements are essentially true when the totality of findings is considered.

4.2.4 The "Psychological Landscape" Hypothesis

This section aims at explaining the diversity of prefrontal deficits observed in aggression. The topographic and hierarchical organization of the prefrontal cortex is used as an explanatory principle. A complementary "Neural Network" hypothesis is presented in the next section.

4.2.4.1 "Psychological Homunculi" in the Prefrontal Cortex

The metaphor refers to the sensory and motor homunculi that visualize the organization of primary sensory and motor areas, which are topographically precise representations of the body. It occurs that the prefrontal cortex is organized in a somewhat similar way: it represents psychological faculties in a mosaic-like, topographically structured, and hierarchically organized fashion. This feature of the prefrontal cortex is highly relevant for aggression control and may explain a great deal of the discrepancies outlined in the previous sections. In the following we will briefly overview the proposed cognitive and emotional maps of the prefrontal cortex. The integration of these findings with those obtained in aggression research and the hypothesis per se will be described in separate sections.

Human Studies

The prefrontal cortex includes motor association areas that serve as interfaces between cognitive and motor networks and as such have important roles in behavioral integration. Therefore, the "psychological topography" of the prefrontal cortex will be started with this "cognitive-motor" level of integration and will be continued with non-motor prefrontal regions. Figure 4.8a summarizes the topographic organization of functions within motor association cortices based on Picard and Strick (2001). Others may attribute different roles to the topographic units and may delimitate them in slightly different ways (see, e.g., Hoshi and Tanji 2007; Kollias et al. 2001; Rushworth et al. 2004). The latter author for instance suggested that the pre-supplementary motor area (the anterior part of the area depicted in Fig. 4.8a) is involved in the selection of action sets, whereas the anterior cingulate cortex relates actions to their consequences and guides decisions about which actions are worth making. Wise et al. (1997) emphasized the role perceptual information that reach the premotor cortex and suggested that this area makes computations not only based on visuomotor but also on proprioceptive, attentional, and other information to produce an output that reflects the selection, preparation, and execution of movements. For the topography of the prefrontal cortex "proper," two models were reproduced in Fig. 4.8b, c. These were based on Badre and D'Esposito (2009) and Petrides (2005), respectively. Albeit the two models appear quite different at the first sight, a closer look can detect similarities that are

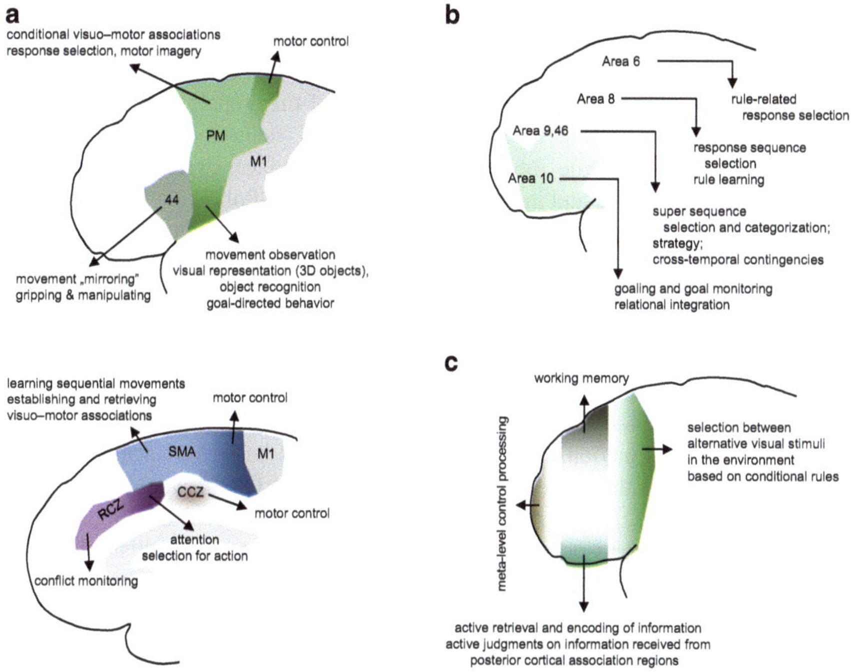

Fig. 4.8 *Models for the topographic and hierarchical organization of the prefrontal cortex.* (**a**) motor association areas (based on Picard and Strick 2001); (**b**, **c**) other areas of the prefrontal cortex (based on Badre and D'Esposito 2009 and Petrides 2005, respectively). *CCZ* caudal cingulate zone; *M1* primary motor cortex; *PM* premotor cortex; *RCZa* anterior rostral cingulate zone; *RCZp* posterior rostral cingulate zone; *SMA* supplementary motor area. Numbers depict Brodmann areas

somewhat covered by the distinct phraseology. More elaborated models also exist; Goldman-Rakic (1996) for instance proposed a detailed prefrontal topography for working memory alone. We would like to emphasize what might not be transparent from the generalizing models reproduced (and partly simplified) in Fig. 4.8. The experimental work summarized in such models paid great attention to the precise location and extent of the functional units of the prefrontal cortex. They were highly attentive not only to the topographic details but also to the overlaps and non-overlaps reported by concurrent studies.

The "emotional maps" of the prefrontal cortex may not be as elaborate as cognitive ones, but the available information does support the notion that emotions are also parceled within this brain region. For example, the orbitofrontal cortex was specifically involved in impulsivity in all three rodents, monkeys, and humans (Matsuo et al. 2009; Rosell et al. 2010; Sellitto et al. 2011; Soloff et al. 2003; Winstanley et al. 2010). Moreover, different aspects of impulsivity may be controlled by different prefrontal regions, the orbitofrontal cortex being involved in motor impulsivity in the first place (Antonucci et al. 2006; Fineberg et al. 2010;

Gansler et al. 2011). The anterior cingulate by contrast was involved in the emotional processing of pain, the same area being linked to motor centers responsible for expressing affect and executing goal-directed behavior (Bie et al. 2011; Shackman et al. 2011). A quantitative voxel-based meta-analysis of 105 studies showed that the processing of emotional faces (vs. neutral ones) activates the medial frontal gyrus, while the processing of angry faces was associated with activations in the middle and inferior frontal gyri (Fusar-Poli et al. 2009).

One has to admit that the existing parcellations of the prefrontal cortex may be provisional, and the location of particular functions may not be sufficiently understood at the time being. The terminology may be author specific too. We are not yet ready to draw a cognitive/emotional homunculus around the prefrontal cortex. Besides uncertainties, however, there is a general consensus on three basic issues: (1) various cognitive functions have specific localizations within the prefrontal cortex, (2) the functions and their neural underpinnings are hierarchically organized, and (3) the functioning of the whole system is based on an intricate networking that ensures the proper handling of information. On the whole, the level of integration increases in the rostral direction: the lowest level of integration is represented by motor association areas, while the highest-level decisions appear to be taken in the frontopolar area. In addition to the rostrocaudal gradient of integration, there also is a dorsoventral dissociation of functions; moreover, differences between the medial and lateral walls of the prefrontal cortex suggest mediolateral topography too. The system is tridimensional.

Animal Studies

The studies underlying the models presented above were based on work performed partly in monkeys partly in humans. Findings were quite congruent and were usually discussed together. Therefore, we will not discuss monkey findings separately. As it regards rodents, the prefrontal topography of behavioral functions is rarely studied in a fashion similar to that seen in monkey and human studies. Nevertheless, a certain kind of topographic organization was found for information processing, learning, and stress responses (Duncan et al. 1993; Gabbott et al. 2003; Stivers et al. 1988; Woolley et al. 2013). Noteworthy, the rat prefrontal cortex, similar to the human one, is involved in motor control.

More detailed maps were established for prefrontal projections. Such studies show a tight topographic organization of prefrontal projections to the amygdala, cerebellum, deep mesencephalic nucleus, hypothalamus, lateral habenula, medullary reticular formation, pontine reticular formation, retrosplenial complex, striatum, superior colliculus, thalamus, and ventral pallidum, i.e., to most of the structures studied so far (Cassell and Wright 1986; Delgado-Martínez and Vives 1993; Floyd et al. 2001; Groenewegen 1988; Kim and Lee 2012; Mailly et al. 2013; Ray and Price 1992; Sesack et al. 1989; Shibata and Naito 2008; Verwer et al. 1996; Watson et al. 2009). In addition, connections between prefrontal subareas and other cortical areas are also topographically organized (Datiche and Cattarelli 1996;

Gabbott et al. 2003). The dopaminergic and serotonergic innervation of the prefrontal cortex shows topographical organization as well (Descarries et al. 1987; Meloni et al. 2008). Taken together, these studies show that the rodent prefrontal cortex is topographically organized; unfortunately no information is available on hierarchical organization, while data on behavioral topography are sparse.

Psychological Homunculi and Aggression

There are many psychological features that engender aggression; these were listed and referenced in earlier sections. Such features can be grouped into two major domains: an emotional and a cognitive one. The cognitive domain includes difficulties in recognizing the valence of social situations (e.g., the emotions and intentions of social partners), the inability to assess behavioral options, as well as the failure to weigh the immediate and long-term consequences of behaviors (lack of insight and foresight). The emotional domain has two "branches." Excessive irritability and impulsiveness (hyperarousal/deficient inhibitory control) may engender reactive/emotional aggression, while psychological stress resilience (hypoarousal) and indifference to consequences for others and self (e.g., lack of empathy and insensitivity to punishment and reward, respectively) may engender proactive-instrumental aggression. Both cognitive and emotional deficits have developmental aspects and may result from inappropriate emotional and social learning. Taken together, these features may lead to faulty decision making and deficient norm compliance and may also be perceived as signs of deficient moral reasoning.

The two domains are tightly related for which "normality" in one domain may be overwhelmed by deficiencies in the other. For example, one may not lack insight and foresight under normal conditions, but may lose these abilities under social pressure if deficient in the emotional domain. Convenient examples are perpetrators of the reactive/emotional type. Such people often show remorse but cannot resist internal impulses in "hot" situations. Conversely, complete control over emotions does not prevent somebody from being aggressive if deficient in the cognitive domain. For example, perpetrators of the proactive-instrumental type tend to pursue immediate rewards even if these entailed long-term risks. In other terms, aggressiveness does not necessitate the presence of all the features that are linked to it; moreover, one single (e.g., impulsivity) or a few psychological deficiencies (e.g., impulsivity + irritability) are enough to engender aggression under specific circumstances (e.g., social challenge). In other terms, the psychological features reviewed above constitute an interdependent emotional/cognitive system that—under certain conditions at least—may result in inappropriate behavior even if only a fraction of its "individual elements" went wrong. We stress the high context dependency of psychological malfunctioning for aggression. The aggression-related consequences of psychological deficits need triggering events and particular conditions to be expressed.

On the neurobiological side, decisions are devised by hierarchically organized systems that rely on an intricate network of information exchange between its building blocks, i.e., the various topographic units of the prefrontal cortex. The consequence of the complexity of the network and its hierarchical organization is that the system as a whole works well only if its units worked well. In other terms, faulty functioning in a few topographic units may disturb the whole system. It is highly probable that the nature of the consequences depend on the particularities of the dysfunctional topographic unit. As an example, deficits in movement observation (one of the roles of the ventral premotor cortex; see Fig. 4.8a) may distort decisions in ways other than the misjudgment of information received from lower association regions by the inferior frontal gyrus (see Fig. 4.8c). However, the ultimate outcome may be aggression in both cases, because, as suggested above, many kinds of psychological malfunctions can lead to aggression under particular conditions and under the influence of specific environmental triggers.

We stress here that the functioning of the prefrontal cortex may not be distorted by structural changes or baseline metabolic rates only. As shown above, there are quite many studies where no such changes were observed in the prefrontal cortex (Antonucci et al. 2006; Barkataki et al. 2006; Brambilla et al. 2004; Laakso et al. 2002; Rosell et al. 2010; Rüsch et al. 2003; Sala et al. 2011; Yang et al. 2005). Thus, abnormal manifestations of aggression may develop without gray matter thinning or low basal metabolism when other types of neural malfunction are present. Malfunctions associated with aggression include:

1. Functional asymmetries between the left and right prefrontal cortices that may alter the relative weight of emotional and cognitive information processing (Antonucci et al. 2006; Keune et al. 2012; Lopez et al. 2004).
2. White matter deficits at the level of the prefrontal cortex that may affect communication between the topographic units of the area (Carrasco et al. 2012; Yang et al. 2005).
3. Altered modulatory influences from lower centers. For example, the prefrontal cortex is under the influence of serotonergic inputs from the dorsal and median raphe, which is altered in aggression-related psychopathologies as shown by fenfluramine and mCPP challenge studies (Davidson et al. 2000; New et al. 2002; Siever et al. 1999; Soloff et al. 2005), as well as studies on serotonin availability (Frankle et al. 2005; Payer et al. 2012; Passamonti et al. 2012; Sekine et al. 2006), and serotonin receptor expression profiles (Booij et al. 2010; Meyer et al. 2008; Oquendo et al. 2006; Parsey et al. 2002; Rosell et al. 2010; Witte et al. 2009). Noradrenergic and dopaminergic inputs may also be affected as shown by studies on MAOA and COMT polymorphisms (Passamonti et al. 2006; Shehzad et al. 2012).
4. Altered connectivity with subcortical regions, e.g., the amygdala (Cox et al. 2010; Fulwiler et al. 2012; Marsh et al. 2008; New et al. 2007; Passamonti et al. 2012). In such cases, internal networking may be normal, but the downward communication of decisions to effector systems is deficient.

A Note on Animal Studies

Unfortunately, rodent studies performed so far are not very helpful in understanding the phenomena discussed in this section. Particularly, aggression-related "psychological" functions were poorly studied, probably because the phrase "psychology" sounds strange when applied to a rodent. However, there are a few studies suggesting that this type of translational research is feasible. Recently developed tests for decision making (e.g., rodent gambling tests) and inhibitory control, as well as more classical tests of attention, impulsivity, cognitive set shifting, etc. (Chudasama et al. 2012; Dalley et al. 2002; Kamigaki et al. 2012; Paine et al. 2013; Sagvolden 2006; Stefani et al. 2003) appear suitable for the purpose. Using such methodologies in aggression research—together with the parallel use of neuroscientific tools may reveal how the topographic organization of the rodent prefrontal cortex is altered in models of abnormal aggression and how does this contribute to the likelihood of showing one or another type of aggressive behavior.

The Hypothesis

The hypothesis was visualized in Fig. 4.9a. Based on the topographical and hierarchical representation of psychological faculties within the prefrontal cortex together with its more general roles in personality and temperament (DeYoung et al. 2010; Whittle et al. 2006), we posit that any durable change in prefrontal function affects the functioning of the whole system, which has an impact on the *likelihood* of being involved in aggressive conflicts.

The findings met head-on in the previous sections are in fact not contradictory but complementary and may become even explanatory if a psychological "interface" was inserted between the history of aggression and prefrontal deficits. In our view, asking whether the deficits in cingulate, frontopolar, *or* the superior frontal cortex define the propensity of behaving aggressively is not the right question. The real question is how particular deficits distort personality, temperament, cognition, and/or emotions; how these affect lifestyles; and how the latter increase the likelihood of being engaged in aggressive confrontations. In other terms, associations between prefrontal functions and *aggression histories* may be used to unravel the quadruple relationship between prefrontal mechanisms, lifestyles, psychological faculties, and proneness to aggression. However, this line of reasoning does not reveal the role of the prefrontal cortex in the neural networks that *proximately* control aggressiveness.

4.2.5 *The "Neural Network" Hypothesis*

In contrast to the "Psychological Landscape" hypothesis, prefrontal roles will not be interpreted here in terms of aggression histories but in terms of executive

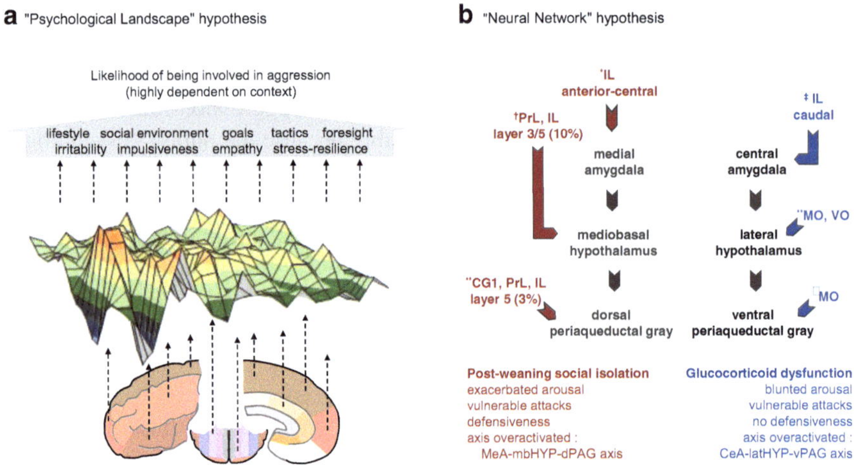

Fig. 4.9 *Intuitive representation of the two models of prefrontal involvement in aggression control*. (**a**) Deficits in particular prefrontal areas (represented by the three views of the prefrontal cortex) affect "psychological landscapes" (represented by the irregular surface), which affect the way in which subjects live their lives and respond to challenges. Life-styles define the likelihood of being involved in social situations where aggression is a viable option, while emotional properties define the likelihood of choosing the option. (**b**) Prefrontal neuronal populations projecting to subcortical structures that control aggression. *Abbreviations*: *IL* infralimbic cortex; *PrL* prelimbic cortex; *CG1* anterior cingulate cortex; *MO* medial orbital cortex; *VO* ventral orbital cortex; *symbols in front of brain areas*, supporting studies (*dagger*, Toth et al. (2010); *double dagger*, Mcdonald et al. (1996); *asterisks*, McDonald et al. (1999); *double asterisks*, Gabbott et al. 2005; *closed circle*, Hoover and Vertes 2011); *percentages in brackets*: the approximate share of the neurons involved (shown where data are available); *blue*, possibly related to glucocorticoid dysfunction-induced (hypoarousal-driven) aggression; *red*, possibly related to early social neglect-induced (hyp*er*arousal-driven) aggression. The characteristics of subjects are indicated in the same colors. *CeA/MeA*, central and medial amygdala; *mbHYP/latHYP*, mediobasal and lateral hypothalamus; dPAG/vPAG, dorsal and ventral periaqueductal gray

functions. The explanatory principle will be the projections of prefrontal neurons to aggression-controlling subcortical structures. The neural mechanisms underlying the execution of aggressive acts are better understood in rodents than in humans; therefore, this section will focus on the former.

The possible involvement of the prefrontal cortex in the aggressive response of rodents was visualized in Fig. 4.9b. Basically we propose that specific cell groups within the prefrontal cortex activate subcortical structures that are involved in the positive modulation and organization of aggressive behavior. The arguments supporting this hypothesis are dual: (1) the execution of aggressive acts is associated with the activation of the prefrontal cortex, and the activation is positively correlated with the level of aggressiveness and (2) there are direct anatomical connections between the prefrontal cortex and subcortical mechanisms involved in the execution of aggressive acts.

The framework makes use of the information summarized in Chap. 3. Current concepts on the control of aggression assume that this behavior is triggered and controlled by a downward stimulatory stream, the components of which are the amygdala, hypothalamus, and periaqueductal gray. Moreover, a distinction was made between a mediodorsal network that controls affective aggression (medial amygdala, mediobasal hypothalamus, dorsal periaqueductal gray), and a lateroventral stream that controls predatory aggression (central amygdala, lateral hypothalamus, and ventral periaqueductal gray). The components of both systems receive direct inputs from the prefrontal cortex.

4.2.5.1 The Medial Amygdala-Mediobasal Hypothalamus-Dorsal Periaqueductal Gray Pathway

The anterior infralimbic cortex contains neurons that project to the medial amygdala; it was hypothesized that this pathway has a role in the control of social and agonistic behavior (McDonald et al. 1999). The infralimbic and prelimbic cortices contain a set of neurons that project to the hypothalamic attack area; it was hypothesized that they activate this region (Toth et al. 2010). These neurons are located in a single cell layer (in layer 3 and/or 5) and are dispersed over the whole rostrocaudal extension of the two prefrontal regions. Finally, neurons located in layer 5 of the cingular, prelimbic, and infralimbic cortices project to and activate the dorsal parts of the periaqueductal gray, where they likely enhance autonomic functions (Gabbott et al. 2005).

We hypothesize that these three prefrontal projections, which target the three main levels of aggression control, enhance activations throughout the pathway. This pathway may be overactivated in abnormal aggression models, where aggression is associated with high emotions (e.g., the postweaning social isolation model).

4.2.5.2 The Central Amygdala-Lateral Hypothalamus-Ventral Periaqueductal Gray Pathway

Neurons located in the *anterior parts* of the infralimbic cortex project to and enhance central amygdala activation (Mcdonald et al. 1996). Neurons of the medial and ventral orbitofrontal cortices project to the lateral hypothalamus (Gabbott et al. 2005). Neurons located in the medial orbitofrontal cortex project to the ventral periaqueductal gray (Hoover and Vertes 2011). We hypothesize that these pathways upregulate the central amygdala-lateral hypothalamus-ventral periaqueductal gray pathway and promote abnormal aggression associated with low emotions (e.g., in the glucocorticoid dysfunction model).

We stress that the relevance of these anatomical connections for aggression control needs further evidence. Figure 4.9b should be perceived as exploratory rather than explanatory. Nevertheless, these connections exist and were relatively convincingly associated with functions that are relevant for aggression control. We also stress that there are a series of other connections between restricted prefrontal regions and subcortical regions that control aggression. However, lengthening the list seems superfluous at this stage, when the roles of such connections all need clear experimental support.

4.2.5.3 The Hypothesis

We hypothesize that the prefrontal cortex is an active component of the pathways that control the execution of aggressive acts. It is not, however, involved as a whole. In fact, not the prefrontal cortex, but certain neuron groups located within this brain area are implicated, each having distinct projection fields.[5] Further research is required to precisely identify prefrontal neuronal populations activated by aggression, their subcortical projections, and their roles in aggression control. In fact, this type of research needs to be begun. We believe, however, that information accumulated so far makes this line of thought worthy of pursuing.

4.2.5.4 Overview

It occurs that the prefrontal cortex controls aggression at two levels. Deficits in particular regions distort the functioning of the whole system, which alters psychological faculties and lifestyles such that the likelihood of being involved in aggression increases. This may result from different anatomical patterns of dysfunctions, because the prefrontal cortex is hierarchically organized. At the same time, the prefrontal cortex contains specific groups of neurons, which have unique projection fields in the subcortical mechanisms involved in aggression control; the activation of these neurons promotes the execution of aggressive acts under particular conditions.

4.3 The Amygdala

The amygdala, similar to all the other brain regions involved in aggression control, has multiple roles. These will be reviewed in the first section of this chapter. Overall, however, this brain region is particularly involved in the control of emotions and the storing of emotionally relevant memory traces, which confers it a specific role in aggression control, namely emotional integration.

---

[5] This active involvement may be direct (stimulatory) or indirect (manifested by the "inhibition of inhibition") as proposed by Potegal (2012). The latter hypothesis, however, does not annul the "active" and execution-bound nature of the involvement.

4.3.1 *Structure and General Roles in Behavior*

4.3.1.1 Anatomy

In the rodent brain, the amygdala is a relatively large structure, which comprises a series of nuclei that are further divided into subnuclei, within which various "parts" are differentiated (Fig. 4.10). Anatomical subdivisions are to a large extent individualized by their connections with other brain areas. For example, the prefrontal cortex is believed to send projections to the basal and central nucleus of the amygdala as well as to the intercalated cells; the medial nucleus receives information from the olfactory bulb; the lateral and basal nucleus is innervated by the hippocampus and the entorhinal cortex, while inputs from the sensory areas of the thalamus and cortex converge onto the lateral amygdala (LeDoux 2007). Outputs of the amygdala nuclei are also specific; e.g., the basal nucleus sends projections to the prefrontal cortex, polymodal associative cortex, and ventral striatum; the efferents of the central nucleus reach modulatory systems (noradrenergic, dopaminergic, cholinergic, and serotonergic nuclei), as well as the periaqueductal gray, hypothalamus, and the dorsal motor nucleus of the vagus, while the lateral nucleus appears to send information to other parts of the amygdala (LeDoux 2007). Internal amygdala communication can be dissected even further. In a recent review, for instance, Pare and Duvarci (2012) described amygdala microcircuits that control fear in terms of subnuclei and specific cell groups within these.

Thus, the amygdala has a complex internal organization and each of its subunits establishes specific contacts with other areas of the brain. While a detailed description of the internal structure and neural connections of the amygdala is not within the scope of this chapter, highlighting its nonuniformity appears quite necessary before discussing its role in aggression. Despite great technological advances, over the last decades, the depth of analysis was not always proportional with anatomical complexities. Nevertheless, sufficient information was gathered over the last few decades to address general issues, e.g., the overall roles of the amygdala in aggression, the roles of particular nuclei, differences between various types of aggression in this respect, etc. However, the amygdala remains a complex structure, and important challenges still lie ahead. The introduction of optogenetic tools holds a great promise in this respect (Anderson 2012).

The amygdala is comparatively much smaller in humans than in rodents, but its structure is no less complex; it appears to be a set of many subnuclei that have different cytoarchitectonic properties, connections, and functions instead of being a homogenous structure with general roles. The subdivisions used in the brain atlas by Mai et al. (1998) are shown in Table 4.4. Based on cytoarchitectonic and connectivity features, the amygdala was lately subdivided into a *superficial complex* (including the anterior amygdaloid area, the amygdalopiriform transition area, the amygdaloid-hippocampal area, and the ventral and posterior cortical nuclei), a *centromedial complex* (including the central nucleus and medial nucleus), as well as a *laterobasal complex* (including the lateral nucleus and the basolateral,

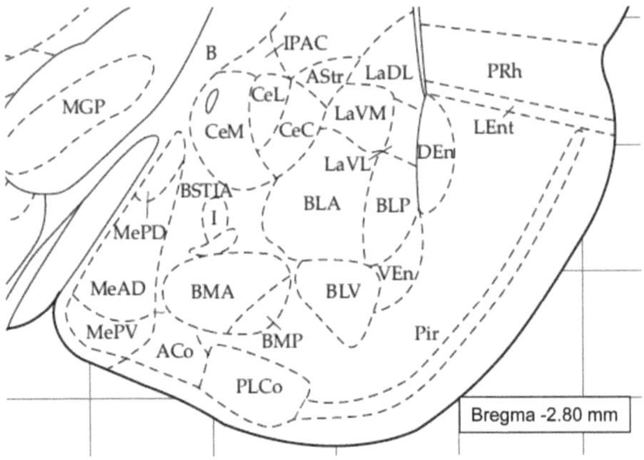

Fig. 4.10 *The structure of the rat amygdala at Bregma −2.8 mm in the atlas of* Paxinos and Watson (1998). This figure illustrates the complexity of amygdala organization in rodents. *Abbreviations*: *ACo* anterior cortical amygdaloid nucleus; *AStr* amygdalostriatal transition area; *BLA* basolateral amygdaloid nucleus, anterior part; *BLP* basolateral amygdaloid nucleus, posterior part; *BLV* basolateral amygdaloid nucleus, ventral part; *BMA* basomedial amygdaloid nucleus, anterior part; *BMP* basomedial amygdaloid nucleus, posterior part; *CeC* central amygdaloid nucleus, capsular part; *CeL* central amygdaloid nucleus, lateral division; *CeM* central amygdaloid nucleus, medial division; *I* intercalated nuclei of the amygdala; *LaDL* lateral amygdaloid nucleus, dorsolateral part; *LaVL* lateral amygdaloid nucleus, ventrolateral part; *LaVM* lateral amygdaloid nucleus, ventromedial part; *MeAD* medial amygdaloid nucleus, anterodorsal part; *MePD* medial amygdaloid nucleus, posterodorsal part; *MePV* medial amygdaloid nucleus, posteroventral part; *PLCo* posterolateral cortical amygdaloid nucleus. Abbreviations of non-amygdalar regions were not explained

basomedial, and paralaminar nuclei) (Amunts et al. 2005; Heimer et al. 1999). As it can be seen, the structure of the three complexes does not perfectly match the system of subdivisions shown in Table 4.4. In addition, connectivity-based and cytoarchitectonic subdivisions do not overlap to a very large extent either (Bzdok et al. 2013). In principle, however, the three complexes (superficial, centromedial, and laterobasal) are parts of rather specific networks. Each complex establishes connections with more than 30 other brain areas (Roy et al. 2009). The connections of particular complexes show very limited overlaps; moreover, the functional directionality of the connection (positive/stimulatory and negative/inhibitory) is often opposite when connections overlap spatially. Left/right differences were also observed. Recently the laterobasal complex was associated with coordinating high-level sensory inputs; the centromedial complex was linked to mediating attentional, vegetative, and motor responses, while the superficial complex emerged as particularly sensitive to olfactory and probably social information processing (Bzdok et al. 2013).

Taken as a whole, brain imaging findings do not reflect the structural and functional complexities of this brain area. Technical development, however, made deeper analyses possible. For instance, Boccardi et al. (2011) and Yang et al. (2009) studied the amygdala in exquisite detail (Fig. 4.11). Although the

Table 4.4 The subdivisions of the human amygdala according to Mai et al. (1998)

Amygdala areas	Nuclei included	Subdivisions
Basolateral nuclear group	Lateral amygdaloid	Dorsal anterior
		Dorsomedial
		Dorsolateral
		Intermediate
		Ventral
	Basolateral amygdaloid	Dorsal
		Intermediate
		Ventrolateral
		Ventromedial
		Paralaminar
	Basomedial amygdaloid	Dorsomedial part
		Centromedial part
		Dorsolateral part
		Ventromedial part
		Ventrolateral part
Central nuclear group	Central amygdaloid	Lateral
		Medial
	Intercalate mass	Not indicated
	Amygdalostriatal transition area	Not indicated
Corticomedial amygdaloid area	Anterior amygdaloid area	Not indicated
	Cortical	Anterior
		Posterior
		Cortico-amygdaloid transition area
		Parahippocampal transition area
Medial nuclear group	Medial amygdaloid	Anterior
		Posterior

amygdala templates used by these authors were not entirely similar, they opened up the possibility of studying the amygdala in detail. It remains to be seen whether the new maps of the amygdala—called probabilistic—are able to add new dimensions to the study of this brain region. These are based on cytoarchitectonic and connectivity data and may dissect functional units better than the more traditional anatomic subdivisions of brain atlases.

4.3.1.2 Function

It is rather difficult to find behavioral functions that are not related to the amygdala. This brain region was considered an important nodal point in the control of anxiety (conditioned and unconditioned), depression, drug dependence and addiction, learning and memory, sexual behavior, pain, etc. (Bellani et al. 2011; Gilpin and Roberto 2012; Guzman-Ramos and Bermudez-Rattoni 2012; Kim and Lee 2012;

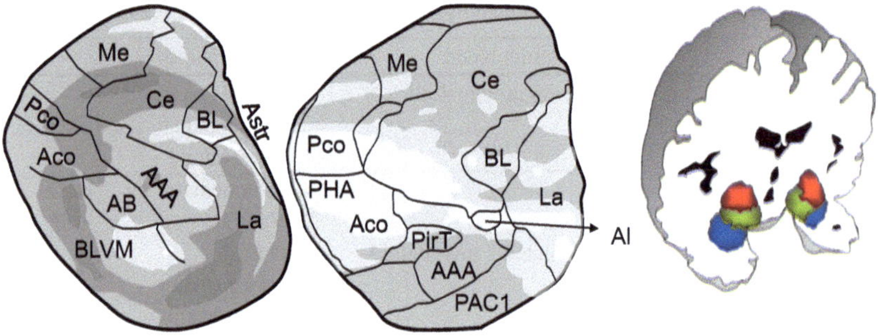

Fig. 4.11 *The structure of the human amygdala on the templates used by* Boccardi et al. (2011, left) *and* Yang et al. (2009, middle) *to evaluate brain imaging data in psychopaths. Light*, volume reduction in psychopathy; *dark*, volume increase in psychopathy. Although the visual perspectives are slightly different, similarities in inferred structures but not in findings are evident. *Abbreviations*: *AAA* anterior amygdaloid area; *AB* accesory basal nuclus; *Aco* anterior cortical n.; *AI* amygdaloid island; *Astr* amygdalostriatal transition area; *BL* basolateral amygdaloid n.; *BLVM* basolateral amygdaloid n. ventromedial part; *Ce* central amygdaloid n.; *La* lateral amygdaloid n.; *Me* medial amygdaloid n.; *PACl* preamygdalar claustrum; *Pco* posterior cortical amygdaloid n.; *PHA* parahippocampal-amygdaloid transition area; *PirT* piriform cortex. For comparison, see the three-dimensional, connectivity-based parcellation of the amygdala on the right (Bzdok et al. 2013). While this parcellation is less detailed, it may be more relevant functionally. *Abbreviations*: *blue*, laterobasal complex, *red*, centromedial complex; *green*, superficial complex. *Note*: all the figures were redrawn by JH

Koob 2003; Newman 1999; Pare and Duvarci 2012; Rasia-Filho et al. 2000; Rouwette et al. 2012; Sigurdsson et al. 2007; Tsoory et al. 2008). The amygdala also plays important roles in aggression as shown below. One can hypothesize that this diversity in functions is related to three characteristics of the amygdala: (1) involvement in the endocrine stress response by direct actions on the hypothalamus–pituitary–adrenal (HPA) axis, (2) involvement in the generation of emotions overall, and (3) the complexity of its organization. Consequently, one has to overcome three problems when trying to understand the role of the amygdala in one particular behavior (e.g., aggression). One has to differentiate endocrine from neural effects and general emotional effects from specific behavioral effects, and one also has to dissect the roles of specific neuron groups within these anatomical structures.

4.3.2 Normal Aggression in Animals

4.3.2.1 Lesion Studies

The destruction of the amygdala decreases aggression in animals. This was shown in lizards, rats, and rhesus and vervet monkeys (Emery et al. 2001; Greenberg et al. 1984; Kling et al. 1970; Meunier et al. 1999; Miczek et al. 1974; for very early studies, see the review by Albert and Walsh 1984). The destruction of discrete amygdala subregions also decreased aggression (*basolateral amygdala, guinea pigs and rats*: Levinson et al. 1980; McGregor and Herbert 1992; *central amygdala, rats*: Karli et al. 1972; *medial amygdala, rats*: Wang et al. 2013; Vochteloo and Koolhaas 1987; Takahashi and Gladstone 1988; *ventromedial amygdala, cats*: Zagrodzka and Fonberg 1977, 1978). Aggression forms decreased by amygdala lesions included predatory (mouse killing in rats), defensive (shock-induced fighting), and rivalry aggressions (resident-intruder conflicts) in males, as well as post-copulatory aggression in females. The available lesion studies are insufficient to evaluate the relationship between various amygdala subregions and particular forms of aggressive behavior, but it is worth noting that a certain kind of specificity was observed in studies where several types of amygdala lesions were studied in parallel. For example, lesions of the corticomedial amygdala did not, while basolateral amygdala destruction did reduce aggression in rats (McGregor and Herbert 1992). The same applies to the destruction of the ventromedial and dorsolateral amygdalae in cats (no effect and reduced aggression, respectively) (Zagrodzka and Fonberg 1978). A few contradictory findings do exist (Blanchard and Takahashi 1988; Machado and Bachevalier 2006; Oakes and Coover 1997); in addition, amygdala lesions reduced aggressiveness in fight-experienced but not in naive rats in one study (Vochteloo and Koolhaas 1987). Taken together, however, the findings overall suggest that the amygdala contributes to the execution of aggressive acts, because its destruction decreases aggression. This conclusion is also supported by the effects of *neonatal* amygdala lesions on aggressiveness shown in later phases of life (Bliss-Moreau et al. 2013; Bauman et al. 2006).

4.3.2.2 Stimulation Studies

The electrical stimulation of the amygdala does not evoke aggression per se in animals, but affects the efficacy of hypothalamic stimulations. A series of studies performed by the group of Siegel show that the stimulation of amygdala subregions has differential—often opposite—consequences (Brutus et al. 1986; Han et al. 1996; Shaikh et al. 1993; Siegel and Brutus 1990). Particularly, the stimulation of the basomedial, cortical, and medial amygdala facilitated affective aggression elicited from the medial hypothalamus, while the basolateral, central, and lateral amygdala inhibited the same electrically evoked behavior. Effects on predatory aggression evoked from the lateral hypothalamus are influenced by the same nuclei in an opposite manner: areas that stimulate affective aggression inhibit predatory aggression and the other way round. Maeda and Maki (1986) i.e., an independent group—replicated and confirmed these findings. Studies in other species are extremely scarce. Potegal et al. (1996a, b) showed that the electrical stimulation of the cortical and medial amygdala reduces attack latencies (stimulates aggression) in Syrian hamsters submitted to the resident-intruder test. This pattern of effects is remarkably similar to that obtained in cats, where the same amygdala regions facilitated the hypothalamic induction of affective aggression.

4.3.2.3 Activation Patterns

Data obtained by the c-Fos technology indicate that the amygdala is activated by aggressive encounters, but its subregions are activated differentially. The most robust effects were noticed in the medial amygdala, which was strongly activated in all the paradigms where it was studied (*California mice, paternal aggression*: Trainor et al. 2008a, b; *hamsters, females*[6]: Joppa et al. 1995; *hamsters, males*: Kollack-Walker and Newman 1995; Pan et al. 2010; *mice, male resident-intruder test*: Duncan et al. 2009; *mice, maternal aggression*: Hasen and Gammie 2005, 2009; *rats, aggression evoked by hypothalamic stimulation*: Halasz et al. 2002b; *rats, resident-intruder test*: Konoshenko et al. 2013; Wang et al. 2013; Veening et al. 2005; Halasz et al. 2002a; Toth et al. 2012; *voles*: Gobrogge et al. 2007; Wang et al. 1997). The medial amygdala was also activated by the mere presence of a potential opponent in adult hamsters (attack priming paradigm) and by play fighting in juvenile hamsters (Cheng et al. 2008; Potegal et al. 1996a, b). Brain imaging studies performed in rats support c-Fos findings and demonstrate a strong activation of the medial amygdala in situations that engender aggressive behavior (Ferris et al. 2008). The available findings suggest that these patterns of medial amygdala activation were selectively related to aggressiveness. The arguments are as follows: (1) aggression levels correlated positively with medial amygdala activation (Duncan et al. 2009; Halasz et al. 2002b, Hasen and Gammie 2005; Trainor et al. 2008a, b); (2) aggression resulted in larger activations than sexual behaviors (Veening et al. 2005); and (3) the "catFISH" technology[7] demonstrated that mating and fighting recruit largely distinct sets of neurons in the medial amygdala, despite similar levels of activation in quantitative terms (Lin et al. 2011). Behavioral selectivity was unclear in other studies (Gobrogge et al. 2007; Joppa et al. 1995; Kollack-Walker and Newman 1995; Konoshenko et al. 2013; Wang et al. 1997). However, one can hypothesize that behavioral selectivity became evident in these studies if the analysis of medial amygdala subregions was as detailed as in the study by Veening et al. (2005) or if the "catFish" technology was available by the time these studies were performed.

The basolateral and cortical nuclei were also activated by the resident-intruder test in all the species studied so far (mice, rats, hamsters, and voles) (Duncan et al. 2009; Gobrogge et al. 2007; Konoshenko et al. 2013; Pan et al. 2010; Veening et al. 2005; Wang et al. 1997) as well as by maternal aggression (Hasen and Gammie 2005, 2006, 2009). An exception is the study by Kollack-Walker and Newman (1995) performed in hamsters where neither the basolateral nor the

---

[6] This was a "regular" resident-intruder test performed in female hamsters; no maternal aggression involved.

[7] The "catFISH" technology exploits the spatiotemporal features of C-Fos expression. When two tests are performed in succession with an appropriate delay (e.g., 1 h), and brains are sampled shortly after the second test, the localization of the c-Fos signal differentiates neurons activated by the first test (signal localized to the nucleus) from those activated by the second test (signal dispersed within the cytoplasm).

cortical nucleus was activated. The reasons of this discrepancy are unknown; noteworthy, however, the medial amygdala was activated in this study as well.

c-Fos studies on the role of the central amygdala (that inhibited affective and promoted predatory aggression in cat stimulation studies) are rather divergent and seem to depend on the type of aggression. In the resident-intruder test, the central amygdala was not or was mildly activated (Cheng et al. 2008; Duncan et al. 2009; Halasz et al. 2002a; Konoshenko et al. 2013; Pan et al. 2010; Potegal et al. 1996a, b; Toth et al. 2012). This suggests that the inhibitory role of central amygdala on stimulation-induced affective aggression "translates" as non-activation in aggressive encounters performed under more naturalistic conditions. Intriguingly, however, the central amygdala was robustly activated in maternal aggression (Hasen and Gammie 2005, 2006, 2009). As shown in Chap. 1, maternal aggression is remarkably similar with certain forms of abnormal aggression in terms of both behavioral features and emotional background. This issue will be discussed in the next chapter.

4.3.2.4 Overview

The three types of approaches provided a consistent picture and clearly show that the amygdala positively modulates aggressive behavior. The power of delimiting the role of amygdala subregions is different in the case of the three approaches. Since the amygdala consists of a tightly interconnected system of subunits, lesions placed in particular subareas likely may disrupt the functioning of the whole system, which explains why the effects of lesions were mildly influenced by their anatomical localization. Stimulation and activation studies suggest, however, that the role of subregions is not equivalent. Aggression is promoted more strongly by the medial amygdala than by any other subarea; by contrast, the role of the central amygdala seems minor, except for maternal aggression, where its role becomes prominent.

4.3.3 Abnormal Aggression in Animals

Abnormal aggression models are associated with altered patterns of amygdala activation as compared to regular resident-intruder tests. The most interesting changes were seen at the level of the central and medial amygdala so far; evidence on other amygdala regions is sparse.

The central amygdala was strongly activated in the pubertal stress and glucocorticoid dysfunction models as well as in mice selected for aggressiveness, and rats selected for low-anxiety behavior (Halasz et al. 2002a, b; Haller et al. 2006; Márquez et al. 2013; Tulogdi et al. 2010; Veenema et al. 2007a, b). In two out of these four models, subjects show both attacks on vulnerable targets and decreased glucocorticoid responses to aggressive encounters (*glucocorticoid dysfunction model*: Haller et al. 2001; *mice selected for aggression*: Haller et al. 2006;

Caramaschi et al. 2008a, b; Veenema et al. 2004). In the pubertal stress model, corticosterone secretion is not decreased overall, but is decreased relative to testosterone secretion (Márquez et al. 2013). Taken together, these findings suggest that the central amygdala is activated when abnormal features of attack are expressed on the background of reduced emotionality. This statement receives support from findings obtained in the maternal aggression model where the central amygdala is strongly activated; as already mentioned above, this form of aggression is also associated with attacks on vulnerable targets and is associated with blunted glucocorticoid stress responses (Gammie et al. 2005; Gammie and Stevenson 2006; Neumann 2001; Parmigiani et al. 1988).

The medial amygdala showed heightened activation in the cocaine, pubertal stress, and postweaning social isolation models as well as in mice selected for aggressiveness and rats selected for low anxiety (Haller et al. 2006; Knyshevski et al. 2005; Márquez et al. 2013; Toth et al. 2012; Veenema et al. 2007a, b). The common characteristic of these models is that subjects showed very high levels of aggression. The activation of the medial amygdala was similar to that seen in controls in the glucocorticoid deficiency model, where abnormal attack is altered in qualitative rather than quantitative means (i.e., attack counts are not increased) (Halasz et al. 2002a, b; Tulogdi et al. 2010). It is also noteworthy that in models where subjects show both abnormal attack targeting and increased attack counts, the overactivation of both the central and medial amygdalae was observed (Haller et al. 2006; Márquez et al. 2013; Veenema et al. 2007a, b).

Unfortunately, aggression-induced amygdala activation patterns were not studied in other models of abnormal aggression. However, neurochemical changes were studied at this level in some models. For example, hamsters chronically treated with anabolic steroids during puberty showed reduced serotonergic innervation and increased GABA synthesis in the medial but not in the central amygdala, and increased vesicular glutamate transporter 2 expression in both the medial and central amygdalae (Carrillo et al. 2011; Grimes and Melloni 2006; Grimes et al. 2003). Reduced serotonin receptor expression was observed also in the mouse variant of the model (the whole amygdala was studied; Ambar and Chiavegatto 2009). In the rat variant of the anabolic steroid model, increased dendritic spine density was observed in the medial amygdala and an amygdala-wide increase of substance P (Cunningham et al. 2007; Hallberg et al. 2000). It is highly probable that these anatomical/neurochemical alterations were relevant for overall amygdala function; consequently, changes in activation levels were detected if studied. Moreover, differences observed in the medial and central amygdala suggest that changes in activation were different in these two amygdala subnuclei.

An interesting aspect of neuronal activations was revealed by two studies where c-Fos expression was studied 24 h after the aggressive encounter. It is worth to note that when studied shortly after encounters, the c-Fos signal identifies neurons that were active in the preceding period, i.e., reveals the brain structures that contributed to the expression of aggression. Twenty-four hours later, c-Fos expression is not relevant in this respect, but may unravel lasting changes in neuronal function. In simple terms, c-Fos labeling studied immediately after the encounter reveals causes (neuronal structures controlling the behavior), while that performed 24 h later may reveal consequences (long-term changes induced by the encounter). In the cocaine model, such long-term changes covered both the medial and central amygdalae, suggesting that the combination of cocaine treatments with aggression (abnormal forms) is consequential for the function of the amygdala on the long run (Knyshevski et al. 2005). Moreover, amygdala subregions were affected differentially in the cocaine model, because the c-Fos signal increased in the medial but decreased in the central amygdala. In the anabolic steroid model, such long-term changes were restricted to the anterior hypothalamus (the attack area in this species), suggesting that the combination of steroid treatment with aggression was consequential for the long-term functioning of this brain region but not for the amygdala (Ricci et al. 2007). It is premature to speculate on the functional significance of these findings, but they indicate that the long-term consequences of aggressive encounters are amygdala-region and abnormal aggression model specific. As repeatedly shown above, model specificity is evident for other mechanisms that control abnormal aggression; studies on the involvement of the amygdala further supports the assumption that there is no single "road" to abnormal aggression; mechanisms depend on etiological factors and behavioral patterns that develop in response.

In conclusion, the data obtained so far suggest that the central amygdala controls qualitative, while the medial amygdala controls quantitative aspects of aggression. Central amygdala activation is associated with attacks on vulnerable targets that are performed on the background of low emotionality, while the activation of the medial amygdala is associated with increased attack counts. Both mechanisms are activated when increased attack counts are associated with abnormal attack targeting.

4.3.4 Normal and Abnormal Aggressions in Humans

Reviews evaluating brain imaging studies on the role of the amygdala—especially those addressing aggressiveness from the point of view of psychopathologies or violent crime—can be divided into three types. Some assert that aggression (especially its abnormal forms) uniquely result from deficient amygdala functioning (Anderson and Kiehl 2012; Raine and Yang 2006). Other reviews differentiate reactive aggression (as seen in spouse abusers and intermittent explosive disorder) and proactive aggression (as seen, e.g., in psychopathy). According to these reviews, the former is associated with upregulated, while the latter with downregulated amygdala functioning (Blair 2010a, b; Crowe and Blair 2008). Finally, other reviews are more cautious with conclusions (Davidson et al. 2000; Dell'Osso et al. 2010). Davidson et al. (2000), for instance, state that "...too much or too little activation of the amygdala may give rise to either excessive negative affect or decreased sensitivity to social cues that regulate emotion, respectively...," by this suggesting that both enhanced and diminished amygdala functions may be perceived as a prelude to violence. We will evaluate below the validity of these assumptions by cross-checking all the methodologies employed so far.

4.3.4.1 Lesion Studies

The findings of Goltz (1892) in dogs and a few similar observations in monkeys (e.g., Klüver and Bucy 1938) prompted Terzian and Ore (1955) to destroy the temporal lobe in humans; they concluded that the phenomenon (called by them the "Kluver and Bucy syndrome") is valid for humans: specific forms of temporal lobe lesions decrease aggressiveness. These disparate findings led Narabayashi et al. (1963) and Narabayashi and Uno (1966) to develop bilateral stereotactic amygdalotomy as a method for the management of severe aggression (the history of this surgical procedure was reviewed by Fountas and Smith 2007). Narabayashi et al. (1963) and Narabayashi and Uno (1966) targeted the lateral aspect of the amygdala; surgeons targeted later the medial (Chitanondh 1966; Hitchcock and Cairns 1973; Mempel et al. 1980), central (Balasubramaniam and Kanaka 1975; Ramamurthi 1988; Sonnen et al. 1976; van Manen and van Veelen 1988), and occasionally to the anteromedial amygdala (Heimburger et al. 1966; Small et al. 1977). The effects of total bilateral amygdalotomy were also reported (Lee et al. 1998).

All the studies mentioned above reported significant ameliorations of aggression after surgery. Albeit the heterogeneity of the subjects and of the outcome measures precludes precise comparisons, improvements were reported to be in the range of 70–80 % (different degrees of improvement, e.g., 33 % and 100 % were occasionally also reported). Where long-term effects were studied, behavioral improvement was maintained for years in a considerable share of subjects (50–100 %). Other symptoms (e.g., post-traumatic personality changes, epilepsy, olfaction-related behavioral malfunctions) were also ameliorated by amygdalotomy (Chitanondh 1966; Heimburger et al. 1966; Hood et al. 1983; Mempel et al. 1980; Small et al. 1977; Sonnen et al. 1976). Interestingly, the lesioned subregion did not seem to predict clinical success; any amygdala lesion decreased aggressiveness substantially; moreover, unilateral amygdalatomy was successful per se in certain cases (for further details, see Mpakopoulou et al. 2008). Overall, the efficacy of amygdalotomy was considered high even if inferior to that of hypothalamotomy, and it was suggested that the two surgeries should be performed in succession if the former did not reach the desired effect (Balasubramaniam and Kanaka 1975; Ramamurthi 1988). This type of surgery was almost abandoned over the last decades mostly for ethical reasons. However its results clearly show that abnormally high levels of aggressiveness require an intact amygdala; moreover, even relatively small lesions are able to abolish severe forms of aggressiveness.

4.3.4.2 Stimulation Studies

Reports on amygdala stimulation are extremely scarce in humans. Faria (2013) mention a few early studies where the electrical stimulation of the amygdala elicited rage in humans. We were unable to access these early studies. Available ones suggest that the stimulation of the amygdala elicits fear and sadness; some electrode locations (especially those located within the left-side amygdala) were able to induce happy feelings (Davis and Whalen 2001; Lanteaume et al. 2007; Meletti et al. 2006). The significance of these findings for the amygdala/aggression relationship is questionable because the overwhelming majority of data were obtained in epileptic patients who were refractory to pharmacotherapy and were prepared for surgery. Even more importantly, amygdala was neither associated with co-stimulation in other brain areas (e.g., hypothalamic centers involved in aggression) nor with challenges of any kind. In addition, it was shown in humans that the consequences of limbic stimulation depend to a large extent on context (Halgren 1982). As such, the aggression-related consequences of amygdala stimulations remain largely unknown in humans.

4.3.4.3 Brain Imaging in "Manifest Aggression" Models

Amygdala function was investigated while subjects were viewing angry faces, or were listening to violent words (both model threats), imagined aggression (e.g., harming others), were induced to feel anger based on autobiographical memory, and were punishing opponents (in the Point Subtraction Aggression and the Unfair Offer Rejection paradigms). All these tasks were associated with the activation of the amygdala (*angry faces, healthy subjects*: Beaver et al. 2008; Hermans et al. 2008; *angry faces, subjects with aggressive tendencies*: Carré et al. 2013; *angry faces, intermittent explosive disorder*: Coccaro et al. 2007a, b; *violent words, spouse abusers*: Lee et al. 2008; *imagined aggression, healthy subjects*: Decety and Porges 2011; *anger induction, healthy subjects*: Fabiansson et al. 2012; *Point Subtraction, borderline personality disorder*: New et al. 2009). Ostensibly, Osumi et al. (2012) reached an opposite conclusion, but their findings allow an alternative interpretation as explained below. These authors studied unfair offer rejection by healthy subjects who showed variable degrees of psychopathic traits.[8] Their findings were dual: subjects high in psychopathic traits were less likely to reject unfair offers (less likely to opt for money loss) and showed decreased amygdala activation when faced with unfair offers. While the authors interpreted this as lower amygdala activation in psychopathy, the data suggest a positive association between the aggressive response (offer rejection) and amygdala activation: if low level of offer rejection was associated with low amygdala activation, then high level of offer rejection was associated with higher amygdala activation. If anything, reduced

[8] The model was described in Chap. 1.

amygdala activation in subjects with psychopathic traits was associated with the avoidance of offer rejection, i.e., of the aggressive response. Interpreted this way, this report also suggests that aggression correlates with increased amygdala activation. Two neutral findings also exist. Amygdala responses to violent words or anger-laden memories did not differentiate healthy from aggressive subjects (Kalnin et al. 2011); angry faces elicited the same amygdala response in borderline personality-disordered and healthy subjects (Minzenberg et al. 2007). The only contradictory finding comes from a study performed by Mathiak and Weber (2006), who reported that violent scenes in violent video games decreased amygdala activity as compared to nonviolent scenes of the same game.

Taken together, "manifest aggression" studies contradict the assumption that aggression per se or abnormal forms of aggression are associated with reduced amygdala activation. On the contrary, the overwhelming majority of studies suggest a positive association between the two variables; a subgroup of studies suggests no association, and only one study is in line with the assumption formulated by Anderson and Kiehl (2012) and Raine and Yang (2006). Notably, however, this single report was on legitimate forms of aggression that was performed by healthy people, because virtual violence is not only accepted but even required by the intrinsic rules of violent video games. The body of evidence reviewed here neither supports nor rejects the hypothesis by Blair (2010a, b) and Crowe and Blair (2008). On one side, subjects likely characterized by high reactive aggression tendencies showed increased amygdala activation in situations that can be related to aggression (Carré et al. 2013; Coccaro et al. 2007a, b; Lee et al. 2008; New et al. 2009). On the other side, however, the same response was seen in healthy subjects (Beaver et al. 2008; Decety and Porgess 2011, Fabiansson et al. 2012; Hermans et al. 2008), and more importantly, the response of subjects likely showing proactive forms of abnormal aggression was also similar (antisocial subjects studied by Spoont et al. 2010; healthy subjects showing psychopathic traits are studied by Osumi et al. 2012). At present stage when data can be considered sparse, one has no choice other than adopting the prudent approach of Davidson et al. (2000) and Dell'Osso et al. (2010).

A Note on Other Types of Findings

When evaluating the role of the amygdala in aggression, authors tend to consider findings that hardly relate to aggressiveness. While angry faces may be perceived as threatening and considered as models of threats, responses to fearful or neutral faces or to merely unpleasant scenes cannot be interpreted in these terms. Responses given to such stimuli are highly relevant for the way in which abnormally aggressive people handle emotions but are not particularly relevant for their aggressive tendencies. The same applies to the prisoner's dilemma game where gain depends on the willingness to cooperate and to situations where the subjects are asked to recognize emotions or to make decisions in situations that have a moral dimension. We do not discuss these studies in detail, because in our view they are relevant to the disorder in general but not to aggressiveness as a particular symptom of these disorders. We note, however, that findings in these studies are mixed, and even if considered relevant for aggression by complex reasoning, they support neither the "amygdala deactivation" nor the "reactive-proactive distinction" hypotheses. The reason is that increased activations were reported in some studies (*conduct disorder*: Herpertz et al. 2008; *borderline personality disorder*: Herpertz et al. 2001; Koenigsberg et al. 2009; Donegan et al. 2003; *psychopathy*: Müller et al. 2003), while decreased activations in others (*conduct disorder*: Sterzer et al. 2005; *conduct disorder with callous-unemotional traits*: Jones et al. 2009; *disruptive behavior with callous-unemotional traits*: Marsh et al. 2008; *psychopathy*: Rilling et al. 2007; Kiehl et al. 2001; Glenn et al. 2009).

4.3.4.4 Brain Imaging in "Aggression History" Studies

The question asked in such studies relates to the link between the general propensity towards aggression and anatomical/functional properties of the amygdala (volume, blood oxygenation, glucose metabolism, etc.). Tacitly, however, they also scrutinize the theoretical underpinnings of amygdalotomy performed to control severe forms of aggression. If aggression histories correlate with downregulated amygdala function, then amygdalotomy should worsen rather than improve their condition.

In most studies, the amygdala was not affected in psychopathologies associated with aggression (*antisocial personality disorder*: Barkataki et al. 2006; *antisocial personality disorder with psychopathic traits*: Bertsch et al. 2013; *borderline personality disorder with antisocial tendencies*: Bertsch et al. 2013; *conduct disorder with callous-unemotional traits*: De Brito et al. 2009; *domestic violence*: George et al. 2004; *psychopathy*: Ermer et al. 2013; de Oliveira-Souza et al. 2008; *violent offenders*: Tiihonen et al. 2008). A few studies, however, did report amygdala deficits (*borderline personality disorder*: Soloff et al. 2008; *conduct disorder*: Fairchild et al. 2013; Sterzer et al. 2007; *psychopathy*: Ermer et al. 2012; Yang et al. 2009). One study reported no quantitative change but an increased left/right asymmetry, which involves downregulation on one side, but upregulation on the other (Raine et al. 1997). There is one study that reported increased amygdala volume in violent offenders (Schiffer et al. 2011). Finally, more complex findings were reported in a subgroup of studies. While studying psychopaths, Boccardi et al. (2011) found decreased volumes in the basolateral amygdala overall, but increased volumes in the ventromedial part of the basolateral amygdala, as well as increased volumes in the central and lateral amygdala. Gopal et al. (2013) found a negative association between the volume of the dorsal amygdala and aggressiveness in psychiatric patients; the association was not valid for the volume of the ventral amygdala. New et al. (2007) was unable to replicate these associations as it regards volumes, but showed a somewhat similar association when glucose metabolism was studied. They found that glucose metabolism did not differentiate healthy subjects from borderline personality-disordered subjects who also suffered from intermittent explosive disorder, but within the latter group, there was a negative association between glucose metabolic rates and aggressiveness. The negative association was valid for both the dorsal and the ventral parts of the amygdala.

As it regards the main issue discussed here (the relationship between amygdala function and aggression), the evidence provided by "aggression history" studies is conflicting and unable to provide a clear picture. By any means, abnormal forms of aggression are not necessarily associated with amygdala deficits, and discrepant data cannot be explained by differentiating reactive and proactive forms of aggression. Regarding the secondary issue addressed, the findings do not clarify how amygdalotomy works. In strictly mechanical terms, it is incomprehensible how the destruction of the amygdala (even unilateral and partial destruction) reduces aggression if the amygdala is not enlarged in aggressive people as a rule. As amygdalotomy was successful in the overwhelming majority of patients, this conclusion means that the effects of amygdalotomy cannot be understood in mechanical terms, i.e., in terms of eliminating a "surplus" of volume. An alternative explanation is provided below.

4.3.5 Overview

A graphical representation of findings was provided in Fig. 4.12. The main conclusions deriving from the studies reviewed above are as follows:

Overall, the studies suggest that the execution of aggressive acts requires an input from the amygdala; the hypothalamus is among the anatomical targets of this input. The supportive arguments are as follows: (1) the hypothalamic induction of attacks is facilitated when the amygdala is co-stimulated and diminished when it is inhibited, (2) the amygdala is activated by aggressive encounters and recent optogenetic evidence suggests that this activation is specific to the execution of aggressive acts, and (3) the amygdala was activated in human subjects who performed tasks that involved feelings or operational responses related to aggressiveness.

Animal studies suggest a "task division" between amygdala subregions: the central amygdala appears to control dangerous forms of attack, especially when these are performed on the background of low emotions. Supporting examples include the cat predatory model, maternal aggression, and a subgroup of abnormal aggression models. The medial amygdala appears to promote attacks on conspecifics overall. Supporting examples include the cat affective aggression model, hypothalamically evoked attacks in rats, and a subgroup of abnormal aggression models.

One can hypothesize that amygdalotomy reduces aggression by disrupting a mechanism that is required for the execution of aggressive acts.

"Aggression history" studies reveal that downregulated amygdala function is *sometimes* associated with a general increase in aggressiveness, especially with abnormal forms. This finding cannot be corroborated with any of the ones briefly evaluated above because it contradicts all. One can hypothesize that such chronic

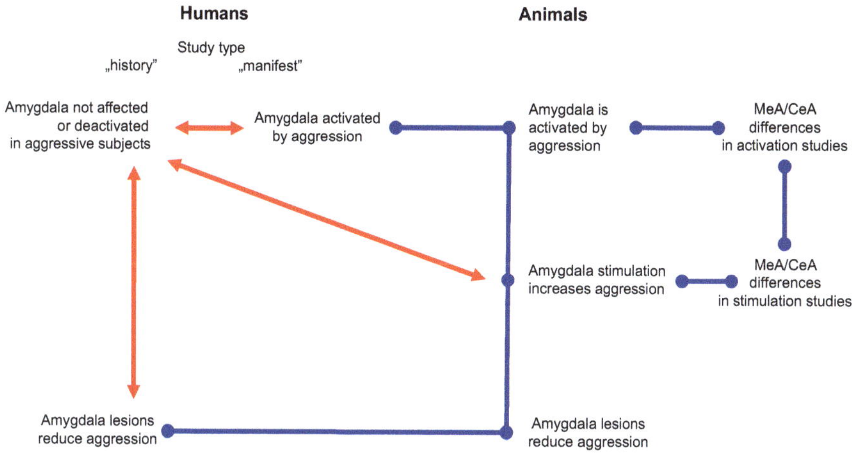

Fig. 4.12 *Simplified representation of findings obtained with the three main types of methodologies in humans and animals. Legend*: *upper row*, activation studies; *middle row*, stimulation studies (no comparative data in humans); *lower row*, lesion studies; "*study type history*", subjects differentated based on their history of aggression; *study type* "*manifest*", subjects differetiated based on aggression-related feelings or behaviors shown while the brain imaging study was performed. Note that all activation studies are manifest studies in animals; *MeA* medial amygdala; *CeA* central amygdala; *blue lines with rounded ends*: compatible findings; *red lines with arrowheads*: contradictory findings

changes in amygdala function affect aggression indirectly by a negative effect on emotions.

4.4 The Periaqueductal Gray

4.4.1 The Role of the Periaqueductal Gray (PAG[9]) in Behavioral Control

The PAG is a tube-shaped midbrain area surrounding the aqueduct, which connects the third and fourth ventricles of the brain. Its relative size is incomparably larger in rodents than in humans. The volume of the PAG in humans (*length*, ~10 mm; *width*, ~6 mm; Satpute et al. 2013) is only *5 times* larger than in the rat (*length*, ~4.5 mm, *width*, ~2.5 mm; Paxinos and Watson 1998), while overall, the human brain (~1,400 g) is about *700 times* larger than the rat brain (~2 g) (see the upper panel of Fig. 4.13). Although never considered, this huge relative difference may be related to the relative weights of "instinctive" and cognitive behaviors in the two

[9] Unlike with other structures, the periaqueductal gray will be abbreviated because of the length of the phrase and because the abbreviation PAG is widely used.

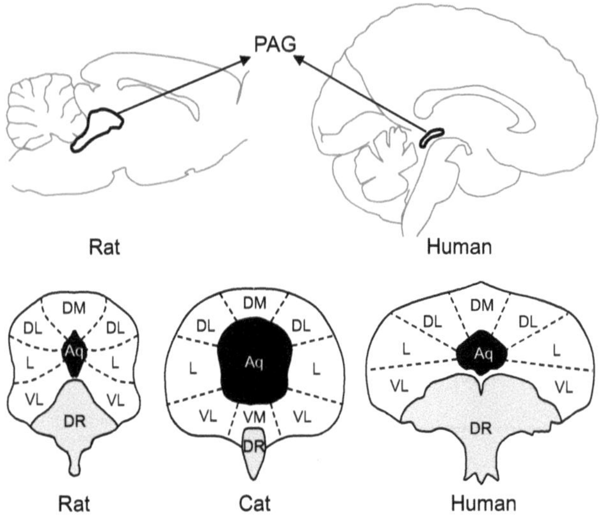

Fig. 4.13 *The relative size (upper panel) and columnar organization of the periaqueductal gray in rats, cats, and humans (lower panel).* Note the large difference between the relative size of the periaqueductal gray in rats and humans. The columnar organization and the relative location of other brain structures is highly similar in the three species. *Aq* aqueduct; *DR* dorsal raphe; *DM, DL, L, VL, VM* the dorsomedial, dorsolateral, lateral, ventrolateral, and ventromedial columns of the periaqueductal gray. Adapted from Paxinos and Watson (1998) (rat); Behbehani (1995) (cat); Carrive and Morgan (2012) and Satpute et al. (2013) (humans)

species, because the PAG appears to control instinctive behaviors prompted by an imminent danger (Bandler et al. 2000; Behbehani 1995; Bandler and Depaulis 1991; Panksepp 2011; Satpute et al. 2013; Subramanian and Holstege 2010; Tracey 2005). In brief, the PAG sets the balance between passive defense (immobility/freezing), active defense (flight), and offense (aggression), i.e., it affects the way in which the organism copes with dangerous situations. In addition to the immobility/flight/fight balance, it also controls associated physiological reactions, like pain-induced analgesia, cardiovascular responses to challenges (including the redistribution of blood flow), activity-related changes in respiration, etc.

The PAG is functionally organized in columns, which structure shows a remarkable cross-species stability (Fig. 4.13, lower panel). In addition, it is also organized rostrocaudally. The PAG seems to have functional units that are arranged along these organizational principles. Figure 4.14a illustrates this by depicting the PAG areas from which the main behavioral responses to environmental challenges (freezing, flight, and fight) can be evoked by electrical stimulation.[10] These areas show very limited overlap. The functional organization of the human PAG is poorly understood because of technical difficulties caused by the aqueduct, which

[10] The cat was used as an example, because fight cannot be easily evoked from the PAG of the rat, while data in humans are sparse as shown below.

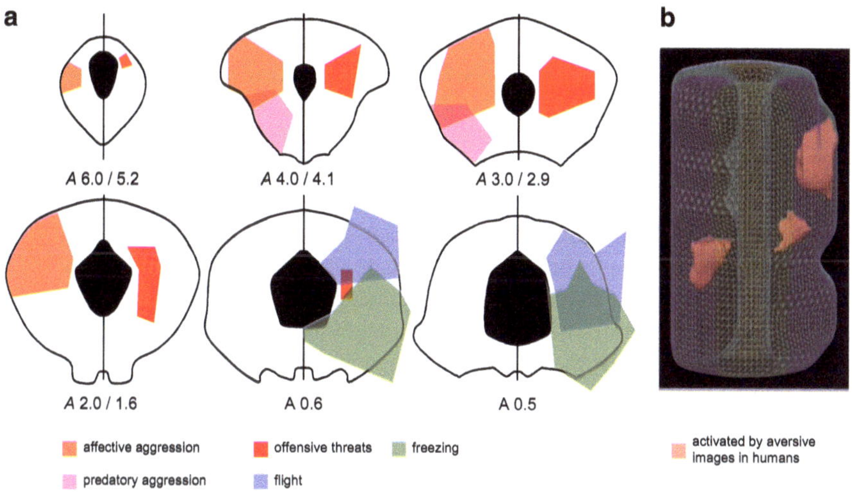

Fig. 4.14 *The columnar and rostrocaudal organization of behavior control in the PAG.* (**a**) *Offense and defense in cats*. The left-hand side of each schematic was adapted from Siegel et al. (1999). The right-hand side was adapted from Bandler and Depaulis (1991). Figures provided by these authors were schematically redrawn on common brain templates. *A*, anterio-posterior distances (if not matching exactly, the figures of both studies were shown). *Color codes*, behaviors elicited by electrical stimulation. (**b**) Three-dimensional reconstruction of PAG areas activated by viewing aversive images in humans (from Satpute et al. 2013)

interferes with the detection of the signals analyzed by brain imaging techniques. Recent advances, however, seem to allow a deeper analysis of PAG structure and activation patterns in humans (Satpute et al. 2013), which suggests that a detailed understanding of the functional organization of the human PAG is not very far away (also see Fig. 4.14b).

Despite its importance in both the control of behavior and associated physiological responses, the PAG received little attention in aggression research so far. This is especially blatant when compared with other areas, e.g., research on defensive responses which phrase usually covers freezing and flight in animals. In humans, the main focus was on anxiety, depression, and pain. Below we overview the evidence on aggression in two short sections, one focusing on animals, while the other focusing on humans.

4.4.1.1 A Note on Offense and Defense

Offense and defense are often viewed as two sides of the same coin, and consequently, the two types of behaviors are often considered together. However, evidence supporting the "one coin" hypothesis is weak. In animals, the main defensive behaviors associated with PAG function (e.g., freezing and flight) are clearly manifestations of fear and, as such, are outside the scope of the present work. Unquestionably, fear may lead to aggression, and certain forms of aggression (e.g., defensive rage in cats) may include a component of fear. Nevertheless, a feeling that leads to a behavior should be not equaled with the behavior per se. Therefore, freezing and flight—tightly controlled by the PAG will not be discussed here in detail. Regarding these behaviors, the reader may wish to read excellent reviews by Adams (2006), Bittencourt et al. (2005), and Canteras et al. (2012).

Although not discussed in detail, the data suggest that the balance between defensive and aggressive behaviors (and their vegetative concomitants) is integrated at the level of the PAG, similar to the integration of cognition and aggression at the level of the prefrontal cortex, the integration of emotion and aggression within the amygdala, and the integration of aggression and endocrine control at the level of the hypothalamus.

4.4.2 *The Role of the PAG in Animal Aggression*

4.4.2.1 Stimulation Studies in Cats

The role of the periaqueductal gray (PAG) in aggression seems to be discovered by Adams (1968) who reported that cells located in this brain area express action potentials during intraspecific fighting in cats and the response is specific to this behavior. Moreover, he showed that the electrical stimulation of the same locations induced hissing, which is part of the affective aggression repertoire of this species. A few years later, Bandler and Flynn (1974) showed that the electrical stimulation of the dorsal PAG elicited predatory attacks in cats. Interestingly, these authors found that this behavior is controlled by a thalamo-dorsal PAG connection, an aspect that received little attention later on. Subsequent studies by Bandler (1977) showed that predatory attacks elicited from the lateral hypothalamus and the PAG were behaviorally very similar, but PAG-induced predatory attacks occurred at considerably lower current intensities than those induced by the stimulation of the lateral hypothalamus. Intriguingly, however, predatory attack sites were located within the ventromedial PAG this time. Somewhat later, affective aggression was also induced from the PAG; the effects of electrical stimulations and glutamic acid injections were similar, suggesting that behavioral effects were not secondary to the stimulation of passing fibers (Bandler 1982). These early studies have the merit of discovering the role of the PAG in both affective and predatory aggressions; the functional organization of the PAG and the involvement of its subregions in behavioral control became clear much later (Bandler and Shipley 1994).

The aggression-related effects of the PAG electrical stimulation were studied subsequently by the group of Siegel, who published many studies on the issue. Beyond delimiting areas involved in affective and predatory attacks with precision (Fig. 4.14a) and showing that electrical and chemical stimulations have similar consequences for both types of aggression, these authors established the connections and the neurochemical nature of connections that subserve PAG mechanisms of attack behavior (Bhatt and Siegel 2006; Bhatt et al. 2005, 2008; Fuchs et al. 1985; Gregg and Siegel 2001, 2003; Hassanain et al. 2003, 2005; Pott et al. 1987; Shaikh et al. 1987, 1991; Zalcman and Siegel 2006). These findings were summarized in several reviews (Siegel and Pott 1988; Gregg and Siegel 2001; Siegel et al. 1997, 1999, 2007). The main findings of this group are as follows.

PAG regions involved in affective aggression (dorsolateral aspect of the rostral half of the PAG) appear to be activated by inputs received from the anterior medial hypothalamus and activate three categories of mechanisms involved in the execution of attacks. One category of outputs is involved in the control of autonomic functions; these appear to be mediated by the locus coeruleus and solitary nucleus and are believed to directly or indirectly target intermediolateral cell column of the thoracic and lumbar spinal cord. The second category of outputs is believed to control motor aspects of affective aggression; they are mediated by the motor nuclei of the trigeminal and facial cranial nerves (responsible for jaw opening and vocalizations associated with affective aggression) and the reticular formation, which target the alpha and gamma motor neurons of the cervical cord (responsible for striking with upper limbs, a component of affective aggression). The third category of outputs targets the rostrocaudal extent of the medial hypothalamus and is perceived as a positive feedback mechanism responsible for the prolongation of affective aggression (see Fig. 3.1). Notably, "backward" projections from the dorsal PAG to "hypothalamic affective attack sites" were noticed rather early (Smith and Flynn 1980). Taken together, these considerations suggest that the periaqueductal gray has a major organizing role in the case of affective aggression.

PAG regions from where predatory attacks can be elicited by electrical stimulation appear to play a more indirect role. This behavior seems to be controlled almost entirely by the perifornical lateral hypothalamus of the cat, particularly by its projections to the tegmental fields, locus coeruleus, and motor and sensory nuclei of the trigeminal complex. It was suggested that predatory attacks can be elicited from the ventrolateral aspect of the PAG by virtue of its projections to the perifornical lateral hypothalamus; thus, the effects of the former are mediated by the basic hypothalamic mechanism involved in the control of predatory aggression.

All the major neurotransmitter systems appear to be involved in the circuits described above (from acetylcholine to vasopressin); in addition, cytokines may also have a role. These issues will be addressed in the next section.

4.4.2.2 Stimulation Studies in Rats

The role of the PAG in intraspecific aggression in rats does not seem to be as clear as in cats. Out of the 112 electrode placements studied by Mos et al. (1982), only 10 resulted in attacks on a partner rat after stimulation; the rest induced locomotion, vocalization, jump, or "alarm-like reactions." Nevertheless, this study also showed that PAG stimulations decreased the electric thresholds for attacks that were evoked from the mediobasal hypothalamus (the hypothalamic attack area), suggesting that the PAG does have a role in the control of intraspecific attacks in this species. A subsequent study by the same group confirmed these findings by employing a different approach (Mos et al. 1983). Large PAG lesions increased the electric thresholds for hypothalamic attack and reduced aggressiveness in resident-intruder conflicts. However, lesioned rats were fully capable of delivering attacks in both paradigms, and effects on resident-intruder conflicts were transient only. Remarkably, cat/rat differences in the effects of PAG stimulation were not associated with differences in connections; similar to cats, hypothalamic areas involved in intraspecific attack specifically targeted the dorsal aspects of the PAG in the rat (Roeling et al. 1994).

It was hypothesized that the differential involvement of the PAG in cat and rat aggression results from differences in the ecological positions and particularities of the two species (Siegel et al. 1999). An unaddressed possibility is that the stimulation models employed in the case of the two species are rather different. While the delivery of attacks was never investigated in the cat affective aggression model (defensive rage), the delivery of attack is the central behavior studied in the rat hypothalamic aggression model. Consequently, one can hypothesize that the two models converge regarding intraspecific aggression-related affective responses and at least do not diverge regarding aggressive responses. The studies performed by the Siegel group did not address attack behavior per se, because opponents are not offered to stimulated cats (see above); what these studies show is that the PAG is involved in behaviors that can be perceived as the first phase of affective aggression that mostly consist of offensive threats. The studies performed by Mos et al. (1982, 1983)—in fact the Kruk group—show that similar responses can readily be obtained in rats by the stimulation of the PAG (vocalizations and "alarm responses"). The studies by the Siegel group are not particularly relevant from the point of view of the actual execution of intraspecific attacks; the studies by the Kruk group show that such attacks are influenced but not directly controlled by the PAG. The two sets of findings combined may show that the PAG readily controls intraspecific threat behaviors and contributes to the expression of, but does not directly control, intraspecific attacks.

4.4.2.3 PAG Activation Patterns in Aggression

In contrast to electrophysiological studies where most studies were done in cats, brain activation patterns were studied mostly in rodents. The techniques employed include immunocytochemical staining for the c-Fos signal, pCREB, and Egr-1; the single study in monkeys used the [(18)F]-fluorodeoxyglucose technique. Before entering into details, two issues need to be specified. The tools used to investigate brain activation patterns may have a considerable impact on the findings. For instance, Gammie and Nelson (2001) studied brain activation patterns by staining for both c-Fos and pCREB and found that the sites of activations did not entirely overlap, but provided complementary information on neural regions active during maternal aggression in mice. This was especially blatant for the PAG, where c-Fos did not, while pCREB staining did evidence an aggression-induced increase. While this information needs to be kept in mind, it is also noteworthy that (1) overlapping sets of authors did find PAG activations with the c-Fos technology in other publications on the same paradigm and (2) it is unknown whether a similar method specificity of findings is valid for males, other species, and other aggression paradigms because similar comparisons are lacking. The second important issue relates to the depth of analysis. In some studies, comparisons were made at the level of the whole PAG; in others the columnar, while in yet others the rostrocaudal organization of the PAG was in the focus of comparisons. Moreover, different phrases denoted overlapping neural structures in some studies. For example, the phrase "dorsal PAG" stood in some studies for a region that covered the dorsomedial PAG and dorsal parts of the dorsolateral PAG, while the phrase "lateral PAG" covered lateral parts of the dorsolateral PAG and the lateral PAG per se. In addition, only one particular PAG region was studied in some cases, which means that changes in other parts of the PAG remained unstudied. Thus, brain activation studies remain spatially imprecise.

Most information on PAG activation patterns was obtained in the resident-intruder and the maternal aggression tests. Taken as a whole, c-Fos counts in the PAG were increased several folds during resident-intruder tests (*rats*: Halasz et al. 2002a, b; Toth et al. 2012; Tulogdi et al. 2010; *male hamsters*: Kollack-Walker and Newman, 1995). In male hamsters, the PAG was among the very few brain areas where aggression elicited a larger activation than sexual behavior. Increased aggressive displays (cage shaking, lunge, ho! threat, stare threat, and head bob) correlated with higher PAG activation when male macaques were exposed to the sight of their females mounted by unfamiliar males, confirming the role of the PAG in monkey aggression (Rilling et al. 2004). Interestingly, regular resident-intruder tests performed in female hamsters did not activate the PAG (Joppa et al. 1995). It is worth to mention, however, that only the dorsal PAG was studied in this publication; in addition, the PAG was studied at one rostrocaudal level only. In terms of internal organization, all PAG columns were activated where these were studied (Toth et al. 2012; Tulogdi et al. 2010). Unfortunately, the columns cannot be compared quantitatively because their size is largely different. As a consequence, absolute counts (all c-Fos counts within a column) and activation densities (c-Fos counts per unit surface) result in different orders of activation levels. When the activation densities were considered, the largest activation was seen in the dorsomedial and dorsolateral PAG, two regions from where attacks can be elicited by electrical stimulation in the cat (unpublished findings in the studies by Toth et al. 2012 and Tulogdi et al. 2010).

The PAG was activated by maternal aggression as well. A BOLD study showed that this activation is prompted by the mere presence of danger to pups (the sight of an unfamiliar male together with the sight of pups) (Nephew et al. 2009). The execution of maternal aggressive acts was also associated with an overall activation of the PAG (Gammie and Nelson 2001; Hasen and Gammie 2006; Lee and Gammie 2007). It occurs that the caudal PAG, especially its ventrolateral section, plays a larger role in maternal aggression than other PAG regions (Gammie and Nelson 2001; Lee and Gammie 2007).

In the only study that addressed predatory aggression (insect predation in rats), rostral regions of the lateral PAG were activated (Comoli et al. 2003).

Taken together, these findings show that the PAG is activated by all three resident-intruder, maternal aggression, and predatory aggression in animals. A few observations suggest that the activation is stronger in the dorsal PAG of males which fight intruders, in ventral aspects of the caudal PAG in lactating mice which fight male intruders, and in the rostral lateral PAG in rats which hunt insects. Thus, immunohistochemical findings confirm the role of the PAG in aggression and also support the notion that the columnar and rostrocaudal organization of the PAG is relevant for aggression control.

4.4.2.4 Other Findings

The neurochemistry of natural forms of aggression is considerably less well known than the neurochemistry of attacks induced by PAG stimulation. A few findings suggest that serotonergic, substance P-ergic, and GABAergic mechanisms within the PAG, as well as the effects of testosterone on this brain region, are relevant for the role played by this brain area in aggression control (De Almeida et al. 2005; Fokidis et al. 2013; Lee and Gammie 2010; Rilling et al. 2004; Shigematsu et al. 2008). These findings are highly similar to those obtained in stimulation studies.

4.4.2.5 Abnormal Aggression

The role of the PAG was investigated in a few abnormal aggression models only. A shift from dorsal to ventral PAG activation was noticed in the glucocorticoid dysfunction model and in mice selected for aggressiveness (Haller et al. 2006; Tulogdi et al. 2010). In the former case, the ventrolateral/dorsomedial activation ratio was increased (i.e., the activation of the ventrolateral PAG column was relatively increased). In mice selected for high aggression (SAL mice) activation levels were lower in the dorsomedial and dorsolateral columns and larger in the ventromedial column as compared to mice selected for low aggressiveness (LAL mice).

These findings suggest that aggression-induced PAG activation patterns in these two models are to a certain extent similar to those seen in models of predatory and maternal aggressions, two models where attacks are also directed towards vulnerable targets and are also associated with low emotionality (Fig. 4.15). Noteworthy, mouse killing in rats also preferentially activates ventral regions of the PAG (Tulogdi et al. in press). We have shown above that the same applies to other brain regions, e.g., the hypothalamus and amygdala.

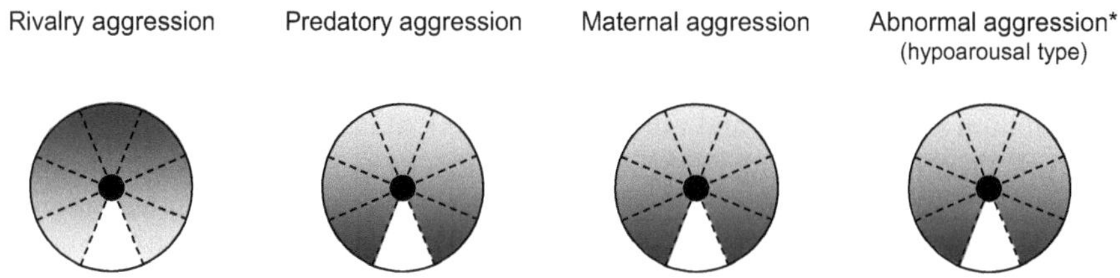

Fig. 4.15 *An interpretation of PAG activation studies in rodents*. The color code indicates regions where the activation is presumably the highest. *Asterisk*, in the glucocorticoid dysfunction model and in mice selected for high aggressiveness. Data suggest that violent forms of aggression executed on the background of low emotionality are associated with a ventral shift of PAG activation

Disparate findings suggest that substance P and opioid neurotransmission in the PAG play a role in the anabolic androgenic model of abnormal aggression (Hallberg et al. 2000; Johansson et al. 2000).

4.4.3 The Role of the PAG in Human Aggression

Data on the role of the PAG in human aggression are very limited. This is due to the technological difficulties explained above; recent advancements in brain imaging techniques (also described above) may rapidly fill these gaps in knowledge.

A recent theoretical paper posits that the PAG particularly its connections with the prefrontal cortex—has an important role in human aggression (Blair 2012). The hypothesis is based on animal findings and human findings obtained in post-traumatic stress disorder. Direct evidence supporting these assumptions was recently published by White et al. (2013). This study employed the so-called social fairness game, a variant of the Ultimatum Game, where subjects are allowed to deliver punishment to (fictional) opponents if these make unfair offers for splitting money.[11] The delivery of punishment was positively associated with the activation of the PAG. The same was true for two other areas, namely the dorsomedial prefrontal cortex and anterior insula. The authors concluded that this form of reactive aggression is under the positive modulatory influence of the PAG and particular regions of the prefrontal cortex.

A recent study by an overlapping set of authors suggested that the PAG is activated by looming threats (Coker-Appiah et al. 2013). Taken together, these human findings suggest that the PAG is activated in early phases of aggressive encounters (e.g., once a threat was perceived) and may contribute to the elaboration of aggressive responses.

[11] This model was described in Chap. 1.

4.4.4 Overview

Stimulation studies in cats and rats show that the PAG is implicated in aggression control. Overall the findings suggest that the PAG has a key role in controlling threat behaviors that precede fights and in ensuring the physiological background necessary to carry out attacks. In addition, the PAG significantly contributes to the execution of attacks; this behavior, however, is unlikely to be controlled by the PAG alone.

PAG activation seems to be rather general in aggression, i.e., it is not restricted to PAG columns from where threats can be triggered by electrical stimulation. One possible reason is that aggression necessitates physiological preparations, e.g., changes in blood flow, respiration, and pain perception. PAG-wide activations may be reflections of vegetative preparations, because this brain region has an important role in these processes. Nevertheless, the activation is not uniform; affective aggression seems to be associated with a preferential activation of dorsal PAG columns, while violent forms of aggression (e.g., those seen in a subgroup of abnormal aggression models) are associated with the activation of the ventral PAG. Considering the localization of predatory attack sites within the PAG, the shift in activation patterns may indicate a co-activation of predatory and rivalry aggression-related PAG sites in these models. This assumption is compatible with changes seen in the hypothalamus and amygdala in the same models.

Chapter 5
The Role of Emerging Techniques in Understanding Aggression

As shown in Chap. 4, the neural control of aggression—despite major advancements—is far from being clear. We review here novel technologies that may significantly contribute to the elucidation of unclear issues and may raise understanding to new levels. Each section starts with a general description of the methodology and its potentials, which is followed by a brief presentation of studies already employing the technology within aggression research.

Each neuroscientific method employed so far advanced knowledge significantly, and their capabilities may not have been exhausted. Qualitatively new information, however, can be obtained by qualitatively new approaches; the ones presented below are likely among those which will shape aggression research in the coming years.

5.1 Optogenetics

Although this technology is relatively new in brain research, its origins can be traced back to the early the 1970s when Oesterhelt and Stoeckenius (1971) showed that bacteria of the genus Halobacterium contain red-colored light-sensitive proteins (opsins) in their cellular membrane. Subsequent studies found that similar proteins are present in the membranes of a series of other microorganisms; in each case, these seemed to be involved in light "recognition," because they elicited the trafficking of ions across the cell membrane when illuminated (Harz and Hegemann 1991; Matsuno-Yagi and Mukohata 1977; Nagel et al. 2002). The technique was introduced in neuroscience by Boyden et al. (2005) who inserted the gene for the naturally occurring algal protein Channelrhodopsin-2, a rapidly gated light-sensitive cation channel into the genome of mouse neurons, which allowed a reliable, millisecond-time scale control of neuronal spiking by blue light. It was also discovered that halorhodopsin, another microbial opsin, is a light-sensitive chloride channel (Schobert and Lanyi 1982); its introduction into neurons by viral vectors and its stimulation with yellow light silence the neurons reversibly also on a

J. Haller, *Neurobiological Bases of Abnormal Aggression and Violent Behaviour*,
DOI 10.1007/978-3-7091-1268-7_5, © Springer-Verlag Wien 2014

millisecond time scale (Han and Boyden 2007). Importantly, the properties of neurons do not appear to be changed by the mere expression of opsins.

The technique shows advantages over electrical stimulation even if employed alone, because it allows both stimulation and silencing; in addition, passing fibers are not affected. Its combination with advanced gene engineering techniques makes the optogenetic technology one of the most powerful tools of neuroscientific research developed so far. Opsin expression can be made selective to neuron types; moreover, the stimulation of axons can dissect the roles of projections.

The optogenetic technology was introduced into behavioral research rather early (Adamantidis et al. 2007; Aravanis et al. 2007). Aggression research benefited so far from it by two research papers and five reviews. Lin et al. (2011) showed that the optogenetic stimulation of hypothalamic neurons induces attacks on both inanimate objects and social partners in mice, while Challis et al. (2013) showed that GABAergic neurons in the raphe mediate the acquisition of avoidance after social defeat. The reviews acclaimed the potentials of the methodology and proposed new directions for aggression research to exploit these potentials (Anderson 2012; Chamero et al. 2012; Nieh et al. 2013; Sternson 2013; Tourino et al. 2013). Thus, the optogenetic technology had no major impact on this field so far. Its potentials, however, are indeed tremendous. We showed above that understanding the mechanisms of aggression control often requires neuron-level analysis of functions, which can be achieved by the optogenetic technology only. Our own experience shows that the introduction of the technology needs expertise in various fields and is not easy; one can confidently state, however, that it constitutes the future in aggression research just like in other fields.

5.2 Epigenetics

The phrase covers processes by which the function of the genome is altered after birth by chromatin remodeling. One of the basic epigenetic mechanisms, DNA methylation, was discovered rather early, but its roles and consequences remained unknown by that time (Gold et al. 1966). In an initial wave of discoveries, epigenetics was associated with viral infections and carcinogenesis (Chu et al. 1977; Rubinstein 1976). However, it became gradually recognized that the phenomenon has more profound roles; it constitutes a basic mechanism that governs vertebrate gene function and cell differentiation (Cooper 1983; Riggs 1975; Razin and Riggs 1980). Nowadays it is well established that epigenetic mechanisms including DNA methylation, as well as histone phosphorylation, acetylation, and methylation, control the transcription machinery, determine cell differentiation, and maintain cell phenotypes over the life cycle (Blomen and Boonstra 2011).

Based on the roles of glucocorticoids in epigenetics (Hofmann et al. 1989), it was proposed that early exposure to stress "predetermines" the development of diseases later in life (including psychopathologies) by affecting epigenetic phenomena (Meaney and Szyf 2005; Seckl and Meaney 2006). Afterwards it was discovered that the phenomenon is more dynamic than previously thought and can be activated and can alter gene expression over the lifespan. For example, chromatin remodeling by histone phosphorylation and acetylation was shown to have a role in the encoding of memories of psychologically stressful events (Reul and Chandramohan 2007). The studies reviewed in these articles in fact revealed a new mechanism by which environmental events affect the functioning of the organism. Nowadays it is widely accepted that major events that affect the individual at any time point of its lifetime elicit lasting changes in gene expression profiles, which is highly consequential for brain function and behavior.

Epigenetic changes—the basic mechanisms which "translate" environmental events into alterations in gene expression—appear highly relevant for understanding aggression-related psychopathologies which are in fact developmental disorders. Moreover, the manipulation of epigenetic phenomena may be important from the point of view of therapy. While this idea did not yet penetrate aggression research, the manipulation of epigenetic processes with therapeutic purposes is an emerging research area in various diseases and psychopathologies (Castren et al. 2012; Gnyszka et al. 2013; Warnault et al. 2013; Zimmermann et al. 2012).

A PubMed search identified 11 studies that addressed the role of epigenetic phenomena in aggressive behavior[1]; 5 were reviews and 6 were experimental studies. The reviews usually interpreted findings obtained by other techniques through the prism of epigenetics and prized the importance and relevance of the phenomenon for aggression research (Cushing and Kramer 2005; Lesch et al. 2012; Tremblay 2008; Veenema 2009, 2012). Interestingly, three of the six experimental studies were done in humans despite the fact that obtaining brain samples is rather problematic in living people. In these reports, peripheral tissues were perceived as markers of central functions. It was found that childhood physical aggression is associated with epigenetic changes in serotonin synthesis (Wang et al. 2012a, b); antisocial behavior in women is linked to childhood sexual abuse by the methylation of the promoter of the serotonin transporter gene (Beach et al. 2011), while BDNF epigenetics was associated with therapeutic responsiveness in borderline personality disorder (Perroud et al. 2013). The main findings of animal studies are as follows: MAO_A epigenetics plays a role in the pubertal stress model of abnormal aggression (Márquez et al. 2013); DNA methyltransferase-1 expression in the amygdala and hippocampus differentiates high novelty-responding and low novelty-responding rats, the former selection line showing high aggressiveness (Simmons et al. 2012); testosterone treatment in the egg affects adult aggressiveness in house sparrows presumably by epigenetic mechanisms (Partecke and Schwabl 2008). Clearly, epigenetic processes are understudied in aggression research. Nevertheless, the general roles played by the former and the type of information the latter needs make the expansion of the field not only likely but also necessary.

[1] Suicidal behavior not considered.

5.3 Functional Pharmacology

Studies on the pharmacology of aggression tacitly or openly assume that the brain control of this behavior is realized on two levels: the behavior per se is controlled by neural circuits, which, however, are under neurochemical influences that impact the circuits as a whole. "Neural mechanistic" approaches operate with concepts like "prefrontal cortex," "amygdala," "hypothalamus," and "periaqueductal gray," while neurochemical approaches deal with "serotonin," "noradrenaline," "dopamine," and "vasopressin." The two approaches seem to run on "roads" that occasionally cross each other but are quite independent overall. While neurochemical approaches negate the role of neurocircuits by no means, there are a large number of studies where, e.g., "serotonin" is perceived as a regulatory mechanism of its own right and where reference to local effects or neuroanatomical entities is minimal. In this section we propose that the amalgamation of the neuroanatomical and neurochemical approaches would markedly improve our understanding of aggressive behavior per se and of its abnormal forms. We will illustrate this by examples.

Example 1—Serotonin Initial enthusiasm regarding the role of serotonin neurotransmission in aggression was hampered by a series of findings that question the existence of a clear-cut antagonistic relationship between the two (Berman et al. 1997; de Boer and Koolhaas 2005; Noblett and Coccaro 2005; Olivier 2004; Takahashi et al. 2011). Discrepant findings were interpreted in terms of phasic and tonic effects, receptor subtypes, concurrent mechanisms, etc. Recent findings suggest, however, that the reasons are deeper; instead of having general effects on aggression, serotonergic neurotransmission appears to affect local mechanisms in highly specific ways. In the alcohol model of escalated aggression, for instance, the expression of 5-HT1B and 5-HT2A receptor mRNAs was halved in the prefrontal cortex, but doubled in the hypothalamus (Chiavegatto et al. 2010). A similar situation occurred in mice submitted to the anabolic steroid model: while the expression of 5-HT6 receptors was markedly reduced in the amygdala, the expression of the same receptor was markedly increased in the hypothalamus (Ambar and Chiavegatto 2009).

Example 2—Vasopressin Recent data suggest that vasopressin neurotransmission has a large role in the control of social behaviors including aggression; moreover, alterations in this neurotransmitter system may underlie the long-term impact of early experiences (Veenema 2009; Veenema and Neumann 2009). However, the interaction is not unidimensional; vasopressin can both promote and inhibit aggression, depending on the brain region where it is released (Veenema et al. 2010). For example, release within the lateral septum correlated positively with aggression, and the local application of a specific vasopressin V1a receptor antagonist lowered it. An opposite interaction was observed in the bed nucleus of the stria terminalis, where vasopressin release correlated negatively with aggression and V1a receptor blockade inhibited this behavior. Similar findings were obtained in hamsters, where vasopressin receptor blockade in the anterior and ventrolateral hypothalamus diminished and facilitated aggression, respectively (Ferris and Delville 1994).

Example 3—Dopamine Dopamine neurotransmission is also critically involved in the execution of aggressive behaviors (Miczek et al. 1994). In line with the findings presented above for serotonin and vasopressin, the interaction between dopamine and aggression appears to be strongly brain area dependent. For example, aggression-induced dopamine release seems to be restricted to discrete dopamine terminal areas, particularly the mesocorticolimbic system (Miczek et al. 2002). Even more interestingly, D2 receptor expression was increased in the anterior hypothalamus but markedly decreased in the ventrolateral hypothalamus in the androgenic steroid model of abnormal aggression (Ricci et al. 2009).

The three examples presented above demonstrate that neurochemical and anatomical mechanisms of aggression control cannot be separated: the overall effects of pharmacological treatments are realized by multiple, often opposite, effects at particular brain levels. The neurochemical aspects of anatomical connections received some attention in the past, especially in the studies published by the group of Siegel (amply discussed above). Even in this case, however, the neurochemical mediators of the main anatomical connections were investigated only. The neurochemical identity of the various inputs (and their subcellular localization) that a particular group of neurons received was never studied. Overall effects (seen, e.g., after systemic or oral treatments with psychoactive drugs) are important from a therapeutic perspective; as such, "global" pharmacology will always be an issue. Nevertheless, the understanding of brain mechanisms requires studies on local effects and on the interactions of these.

5.4 Genetics and Pharmacogenetics

The quest for "aggression genes" was prompted by the strong heritability of aggressiveness (van Oortmerssen et al. 1992; Miles and Carey 1997). The search was successful, probably more successful than expected. The number of genes associated with aggressiveness was surprisingly large irrespective to the technology employed and the subjects studied (36 genes identified by the transgenic technology till 2003, Maxson and Canastar (2003); 335 genes identified by sequence analysis in tame and aggressive foxes, Kukekova et al. (2011); 262 genes associated with seasonal variation in aggression in song sparrows, Mukai et al. (2009); the majority of the marker genes studied in feral rats, de Boer et al. (2003) etc.). A detailed analysis of the genetic underpinnings of aggression is not within the scope of this book. We draw attention on one emerging line of research only, namely, pharmacogenetics.

This phrase refers to genetic differences that affect individual responses to drugs, e.g., genetic differences in receptor structure and ligand metabolism (Klotz 2007). The field is not particularly new; the first paper on the phenomenon was published a long time ago (Evans and Clarke 1961). Although the primary focus was on drugs so far, pharmacogenetic differences are unlikely to affect drug responses alone, because endogenous ligands and therapeutic agents affect similar mechanisms. As such, pharmacogenetic approaches address the very essence of neurochemical communication.

The PubMed search with the terms pharmacogenetics and aggression identified seven relevant papers, out of which four were research studies, while three were reviews (Nesher et al. 2013; Singh et al. 2012; Guillot et al. 1999; Miner et al. 1993; Patel and Barzman 2013; Takahashi et al. 2012; Veenstra-VanderWeele et al. 2000). While such limited evidence seems insufficient to address the issue systematically, the available evidence suggests that (1) abnormal forms of aggression are associated with genetic variability that forecasts significant variations in drug responses and (2) genetic variability in receptor characteristics brings about changes in aggressive behavior. We mention that there are a series of studies which do not bear the flag of pharmacogenetics but are still important in this respect. For example, genetic polymorphisms in the encoding of neurotransmitter receptors affect not only the functioning of aggression-controlling brain mechanisms but also their responses to drug treatments (for recent examples of such studies, see Beitchman et al. 2012; Benis and Hobgood 2011; Butovskaya et al. 2012; Vaillancourt et al. 2012; Zai et al. 2012).

5.5 Overview

One of the main conclusions deriving from the studies reviewed in Chap. 4 is that the interpretation of findings on aggression control is quite often hampered by the lack of detail. Missing pieces in the multidimensional puzzle of aggression control may be supplied by the advanced technologies presented above. The identification of particular groups of neurons that play roles in aggression control and the mapping of their anatomical connections can be done with unprecedented precision by the optogenetic technology and functional pharmacology; the ultimate mechanisms of abnormal aggression and their relationship with environmental factors may be resolved by studies on epigenetic phenomena, while pharmacogenetic studies may reveal how drug effects and neurochemical connections are altered in abnormal aggression. Their combined application will elevate understanding to a qualitatively new level.

Chapter 6
Summary and Synthesis*

6.1 Who's Flying the Plane and What Kind of Plane?

The first metaphor in this title was borrowed from Siegel and Douard (2011) who addressed a philosophical problem raised by neurochemical models of violence. They asked the question whether aggression is biologically deterministic, in which case perpetrators are not responsible for their acts, or it depends on free will, in which case perpetrators are culpable. In our adaptation of the metaphor, both the paraphrased actors and the question are different. The ones who fly the plane are brain areas, while the plane itself is aggression. The question "who's flying the plane" may be reformulated less metaphorically: which brain area controls aggression, to what extent, and in which ways? Which of these brain areas or which of their functions go wrong when subjects show abnormal forms of aggression?

The second metaphoric question refers to the nature of aggression. Defined in the simplest way, aggression is a behavior that aims at inflicting harm. Inflicting harm, however, is not a simple mechanical act. The target moves, works against offensive actions, and may even counterattack. If aggression consisted exclusively of simple, clearly definable behavioral acts (e.g., biting in animals and hitting in people), its neural control would be similarly simple and would have been discovered a long time ago. However, aggression comprises a series of widely different actions, some of which aim at social recognition, others at avoiding harm, yet others at inflicting harm or expressing dominance, and all these need to be coordinated and adapted to environmental conditions (for instance, the actions of opponents) and temporal changes in these. Metaphorically speaking, the "plane of aggression" is a rather complex one.

*We do not intend in this chapter to reiterate the details (and the references to these details) that were amply presented in previous ones. Here, we attempt at providing a synthesis of the finding in broad terms. For most of the supporting evidence, the readers are kindly directed to the earlier chapters.

J. Haller, *Neurobiological Bases of Abnormal Aggression and Violent Behaviour*,
DOI 10.1007/978-3-7091-1268-7_6,

The first section of this chapter is an attempt to corroborate the types of behavioral acts performed during aggression and the types of functions and abilities that particular brain areas possess. Alterations in the system and their possible repercussions for aggressive behavior are discussed in the following chapter.

6.2 Behavior and Brain Function in Animal Aggression

6.2.1 Normal Aggression

In this section we propose that the execution of aggressive acts is initiated by higher-order brain areas, primarily by the prefrontal cortex. This hypothesis challenges the view that aggression is released from, and governed by, mechanisms located in lower-order brain areas (e.g., the hypothalamus and periaqueductal gray) when their inhibition by higher brain structures is suspended. We argue here that the latter two brain areas may indeed put into action neuronal networks that control the execution of fixed action patterns (e.g., the movement sequences of attacks), but they are unable and inappropriate to control aggressive behavior in the context of an ongoing conflict, simply because they have no information on what happens "in the field." The choice and timing of behaviors is more likely controlled by higher-order brain structures where the necessary information converges. The arguments supporting the model are as follows.

1. Many of the behavioral acts performed by rodents during aggressive encounters have a fixed choreography, which is innate.[1] For example, clinch fights in rats are composed by a sequence of movements which starts with bending over the opponent while the animals are positioned in a right angle relative to each other; continues with snatching with teeth the flank of the victim on the side opposite to the attacker, pulling the victim in the air with teeth; and ends by a blow with hind paws that throw the victim away.[2] This behavioral sequence does not depend on social experience; e.g., it is displayed in a highly similar fashion by rats reared in isolation from weaning. Some of the behavioral acts performed during aggressive encounters are composed of simpler movement sequences, but are clearly recognizable as fixed action patterns. This particularity of rodent and more generally animal behavior made the description of ethograms possible.[3] Under particular conditions, a fixed action pattern may be present or absent, may be

[1] Rodents and particularly rats are used here as examples, because a detailed comparison of various species is outside the scope of this chapter. It is worth to note that the principles governing animal behavior are universal and conclusions deriving from these model animals are valid for other species.

[2] This very quick movement sequence is visible on slow motion pictures only.

[3] For the ethogram of rodent aggressive behaviors, see Grant and Mackintosh (1963).

expressed frequently or sporadically, but the basic movement sequences are not altered. If a clinch attack is present, it is always performed in the same way.

2. It is reasonable to assume that innate and fixed action patterns are controlled by specialized neural networks. The whole sequence described above takes less than a second which leaves no room for planning and cognitive control. Given that clinch attacks seem to be innate and fixed action patterns that are executed extremely quickly, we assume that they are evoked by successive discharges of a tightly interconnected set of neuronal structures. Furthermore, it is reasonable to assume that the network is activated by brain areas from where the movement sequence can be evoked by electrical stimulation. For instance, the stimulation of the mediobasal hypothalamus elicits the whole sequence described above.[4]

   While the hypothalamus is part of the network that evokes clinch fights, it is unlikely that it is able alone and solely to control such complex sequences that include movements of the limbs, body, head, and jaws. It is more likely that the hypothalamic attack area plays an initiating role, while the movements of the sequence are governed by lower brain areas and the spinal cord. This neuronal network likely includes offense-related mechanisms located within the PAG, as particular neurons in this brain area innervate centers, which elicit jaw and limb movements (see Sect. 4.4).

   The PAG appears to be important also for another category of behaviors frequently shown during aggressive encounters. Any of the contestants including the intruder may prevail in particular moments of the contest, which makes defense, flight, or immobility necessary for the adversary. These forms of defensive behavior can be elicited by the electrical stimulation of the "defensive" PAG (subregions governing defensive behaviors; see Sect. 4.4); these behaviors appear to be controlled by this brain area in a manner similar to the control of attacks by the hypothalamus.

3. Aggressive encounters consist of intercalated bouts of offense and defense (moreover social investigation and exploration), the expression and succession of which is determined by the momentary needs and constraints of social interaction rather than by predetermined programs. The mediobasal hypothalamus and the PAG may trigger fixed action patterns, but it is highly unlikely that these brain areas are able to adapt the execution of behaviors to the circumstances. The reason is that they do not receive the information necessary to make decisions on what to do and when. Decisions of this type must come from upstream, from brain areas where environmental information converges and is

[4] In fact, stimulation induces one of the following four movement sequences depending on its location and current intensity: attack jumps, clinch attacks, hard bites, and soft bites. Each of these has a fixed choreography as follows. Attack jump: jump towards the opponent in upright position by using hind paws for takeoff, bite on head on landing, and blow with hind paws against the chest or belly of opponents; clinch attacks: see above; hard bites: charge with the head that ends with a bite and a concomitant stamp against the floor with hind paws; soft bites: simple biting (Kruk et al. 1983; Kruk 1991, and personal observations). The same behavioral sequences are seen under natural circumstances.

processed. In addition, the choice of behaviors and their timing needs information on the internal state of the organism and on past experiences and is also governed by emotions. Such information also seems to converge to, and to be processed by, higher-order brain areas. Consequently, decisions regarding the expression and succession of particular action patterns must be taken at levels higher than the mediobasal hypothalamus and the PAG. The most likely brain site of such decisions is the prefrontal cortex. The amygdala may also have a role, but according to current knowledge, the main brain site of decisions is the prefrontal cortex.

4. At the level of this brain area, the processing of information is achieved by communication with lower brain centers (e.g., sensory fields, thalamus, amygdala) and interactions between its subareas. The net result of this intricate network of information exchange seems to result in decisions that are communicated downwards by those prefrontal neurons that project directly to the executive mechanisms contained by lower-order brain areas. Thus, the organizing role of the prefrontal cortex seems to be performed in two steps. The first step includes the processing of information and decision making which involves extended regions of these brain areas. The second step is the communication of decision via a small number of dedicated neurons.

In brief, we propose that behavior in aggressive conflicts is shaped by a cascade of neuronal events, which start by information processing and decision making at the level of the prefrontal cortex (Fig. 6.1). Decisions are communicated to lower brain areas by subgroups of neurons located within the very same brain area. On their turn, lower-order brain areas initiate the activity of specialized neuronal networks that elicit coordinated muscle contractions that sum up as goal-directed behaviors. The whole network seems to oscillate between different modes of function, as episodes of offense (threats and attacks), episodes of defense (defensiveness, flight, immobility), and episodes of "relaxation" (exploration, grooming, social investigation) alternate frequently. Alternation is determined partly by external circumstances (actions by opponents which trigger shifts from offense to defense and back) partly by internal circumstances (for instance, by fatigue which triggers shifts from agonistic behaviors to "relaxation"). These circumstances are continuously monitored by higher-order brain centers, which switch executive mechanisms on and off depending on the situation.

6.2.2 Abnormal Aggression

Our first study on the neural background of glucocorticoid dysfunction-induced abnormal aggression brought about a great surprise: brain areas believed to play an important role in rivalry aggression (the medial amygdala, mediobasal hypothalamus, and periaqueductal gray) were activated by aggressive encounters, but the level of activation was not different from that seen in controls. Changes were seen

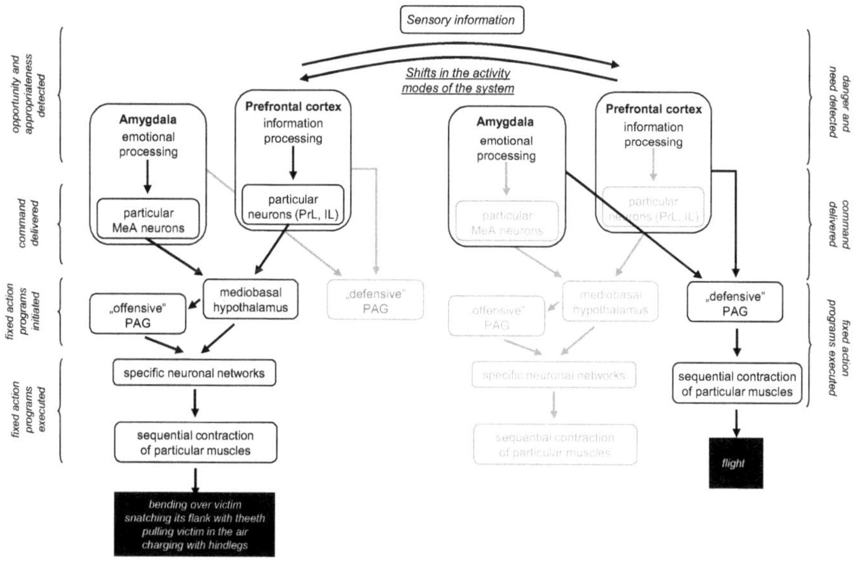

Fig. 6.1 Cascade of neural events leading to clinch attack (*left*) and flight (*right*) during one and the same resident/intruder conflict in rats. Note that periods of offense and defense alternate in both contestants, even if their ratio depends on status. The basic idea is that the choice between alternative behaviors and their timing can be established by higher order brain areas (PFC and amygdala) where the neccesary information converge. The mediobasal hypothalamus and PAG are responsible only for "mechanical" execution and organization (see Sect. 6.1). *Black area names and arrows*, mechanisms momentarily active; *grey area names and arrows*, mechanisms momentarily inactive; *italics*, neural and behavioral events; *"defensive" and "offensive" PAG*, periaqueductal gray mechanisms controlling defense and offense, respectively; *MeA, PrL, and IL*, medial amygdala, prelimbic and infralimbic neurons that project directly to the mediobasal hypothalamus

in brain areas where these were unexpected, e.g., the central amygdala.[5] We experienced a second surprise when other models of abnormal aggression were analyzed with respect to their neural background. Functional brain alterations similar to those observed in the glucocorticoid dysfunction model were seen in a restricted number of models only. Even our second model of abnormal aggression (postweaning social isolation) proved to have different neural underpinnings as evidenced by the c-Fos technology.

The overall conclusion of the studies published so far is that abnormal aggression may result from a variety of brain alterations, which, however, may be grouped into the following three main categories.

[5] Changes in the central amygdala were assumed to be related to increased fear by that time (Halasz et al. 2002a). Subsequent research, however, did not support the "fear hypothesis" (Haller et al. 2004), but suggested that changes in the central amygdala reflect a shift towards a predatory aggression-like neuronal activation patterns (Tulogdi et al. 2010).

1. *Hypoarousal type* models are characterized by downregulated emotional responses to aggression in terms of both glucocorticoid and autonomic changes. In these models, the medial and central amygdala and the mediobasal and the lateral hypothalamus are co-activated; as it regards periaqueductal gray, the center of activation seems to shift ventrally. Similar patterns of emotional and neural responses are typically seen in two types of normal aggression, namely predation and maternal aggression.

   The similarities between this subgroup of abnormal aggression models on one side and predatory and maternal aggressions on the other side suggest that those etiological factors that result in the aforementioned pattern of emotional and neural particularities (e.g. abnormal forms of aggression) do not bring about some kind of "unnatural" ways of neural and emotional functioning, but activate mechanisms that are inappropriate to the situation.
2. *Hyperarousal type* models are associated with increased emotional responses; either or both autonomic and glucocorticoid responses are increased in this type of models. Interestingly, higher emotionality seems to be restricted to social contexts, because subjects show normal or even reduced emotional responses to nonsocial challenges in some models. No brain areas other than those involved in rivalry aggression are activated, but their activation levels are higher than in controls.

   The primacy of emotional and neuronal changes remains to be clarified. High emotions may inherently lead to enhanced activation of brain regions that control behavior during conflict, but it is equally possible that changes in brain function explain both increased emotions and aggression. Either alternative may be supported by the gross spatial co-localization of emotion- and aggression-related mechanisms (e.g., the "amygdala" hosts both). This cannot settle the question of primacy because local (within-area) interactions between the two mechanisms are obscure. Based on the observation that amygdala neurons involved in sexual and aggressive behavior are distinct but spatially intermingled, one can assume that emotion- and aggression-bound mechanisms need to be discerned at neuronal level.
3. In *minor type* models, neural and behavioral changes are subtle. The former are usually restricted to neurochemical changes seen in a few brain regions. Some of them were seen in brain areas that were not analyzed in detail here, e.g., the septum. It occurs that vasopressinergic communication is especially affected in these models.

   "Minor type" models suggest that abnormal forms of aggression may develop even in the absence of major changes in brain function provided that the mechanisms where subtle changes are seen have a major role in aggression control. Noteworthy, however, aggressiveness was less markedly affected in these than in other models of abnormal aggression. Behavioral changes were either restricted to certain periods of development or the number of abnormal features was very limited. This may lead to the conclusion that minor changes in brain function do lead to abnormal forms of aggression, but behavioral changes are also mild.

While it was tempting to classify the available models according to the types of neural changes discussed above, as well as according to their emotional and behavioral features, we refrained from doing so because the models discussed here are in very different stages of characterization. Some are incomplete in terms of behavior description; others are unilateral as it regards the neuroscientific methodologies employed, while emotional background was not studied in others. One cannot exclude that in those models where behavioral changes seemed mild, a deeper analysis would reveal larger effects (e.g., attack contexts and targets may be affected even if attack counts are normal). The same applies to brain areas: the lack of changes in the ones that were investigated does not exclude changes in others. An analysis of emotional responses to social and nonsocial challenges may also considerably affect the categorization of models. Despite the incompleteness of information, one can confidently conclude even at this stage that the neural, emotional, and behavioral changes seen in the three types of models are different and depend on the etiological factor of the abnormal forms of aggression; i.e., there are multiple "roads" leading to dysfunctional aggression.

Emotionally, the "hypoarousal-hyperarousal" types of abnormal aggression are akin to human aggressions identified by the phases "proactive-reactive" or "instrumental-emotional." Behaviorally, however, the applicability of these terms was not verified so far in rodents probably because appropriate testing procedures are not yet available. This seems to be a major drawback, because it hinders direct comparisons with abnormal forms of human aggression. Nevertheless, differential emotionality is so evident in the proactive and reactive human aggression that correspondences may be established based on emotionality alone. This should be done with caution, however, especially because emotionality was incompletely characterized in many models.

6.2.3 The Role of the Prefrontal Cortex

Given the importance attributed to the prefrontal cortex in Sect. 6.2.1, the lack of reference to this brain region in the classification system proposed above may be surprising. The main reason is that the prefrontal cortex requires separate analysis due to the complexity of its roles.

We proposed above that the control of aggression by the prefrontal cortex is exerted in two steps. In the first step, it triggers shifts between offensive, defensive, and "relaxation" modes of action depending on circumstances. In the second step, it sends commands to lower brain centers that are elicited by the decisions made in the first step. Activation profiles observed in models of abnormal aggression were interpreted with reference to this dual role.

1. The properties of the prefrontal cortex may influence the way in which social challenges are handled in general. For instance, increased impulsiveness

developed due to functional alterations in prefrontal cortex may increase the readiness to express aggressive behavior.

2. Behavioral events occurring during aggressive encounters may also have a large impact on prefrontal functioning. There is evidence showing that the behavioral structure of the encounter depends on the model. Márquez et al. (2013), for instance, showed that intruders faced with residents of the pubertal stress model spend around 30–50 % of time in submission, which largely exceeds the levels seen in controls. In the glucocorticoid dysfunction model, intruders are momentarily intimidated by vulnerable attacks, but their submissiveness is not altered overall (Halasz et al. 2008[6]; Miczek et al. 2013). By contrast, rats submitted to the postweaning social isolation model show decreased dominance and increased defensiveness (Toth et al. 2008, 2011). One can assume that the prefrontal cortex needs to adapt to circumstances of this kind and needs to switch from one mode of functioning to another in a model-specific way, which has consequences for the activation patterns observed in the model. For example, the activation of the medial orbital cortex was decreased in the pubertal stress model, was not changed in the glucocorticoid dysfunction model, and was increased in the postweaning social isolation model. This suggests that the activation of this particular prefrontal region depended on dominance relationships rather than on the particularities of the attacks delivered.
3. Finally, prefrontal activation patterns are also determined by the activity of the mechanisms that actually initiate the functioning of lower structures. The prefrontal cortex sends projections to all the brain structures involved in aggression control, including those involved in predatory aggression. One can hypothesize that specific subgroups of neurons that innervate the central amygdala, lateral hypothalamus, and ventral parts of the periaqueductal gray are overactivated in hypoarousal models, while prefrontal neurons innervating the medial amygdala, mediobasal hypothalamus, and dorsal aspects of the periaqueductal gray may be overactivated in hyperarousal models.

6.3 Behavior and Brain Function in Human Aggression

We suggest below that the control of aggression in general and the control of abnormal aggression in particular are governed by very similar principles in rodents and humans. The two research areas followed different routes, for which the making of correspondences is not always simple. For example, rodent data on basal prefrontal functioning in abnormal aggression models are extremely scarce, while similar findings are abundant in humans. By contrast, studying brain function during aggressive encounters is the norm in rodents; similar studies are infrequent in humans. The available information suggests, however, that aggression-related

[6] As judged from the duration of dominance by residents.

brain functions (decision making—execution) as well as their neural substrates (prefrontal cortex—prefrontal cortex, amygdala, hypothalamus, periaqueductal gray) are similar.

6.3.1 The Dual Theory of Aggression

We divided human studies into two categories above. In one of these, brain function is studied while the subjects actually execute aggressive acts. Due to technological constraints, these are inherently mild and are often symbolic. The other category of studies focuses on "real" aggression, but retrospectively: brain function is studied while subjects are cooperative (e.g., obey the rules of the investigation), while aggressiveness is attributed to them based on past criminal actions or diagnoses (the presence of aggression-related psychopathologies). The two types of studies will be evaluated below separately. Normal and abnormal forms will be discussed together but will be differentiated as far as possible. This is due to the general approach of human investigations where the focus is on abnormal forms.

Findings obtained in the first type of studies (called "manifest aggression" studies above) are surprisingly similar to those obtained in animal research: the execution of aggressive acts is associated with the activation of all four the prefrontal cortex, amygdala, hypothalamus, and periaqueductal gray. In the former two areas, activations are subarea specific. At lower levels (hypothalamus, PAG), the techniques employed to date do not allow the analysis of structural subtleties. Nevertheless, the main principles of control seem to be similar, which is not entirely surprising. Aggression is an "ancient" form of competition in the meaning that it is shown by all vertebrates and all lower animals that have sufficiently developed movement abilities and neural systems. From an evolutionary perspective, not similarities but discrepancies would have been surprising; it would be unexpected to find out that humans "invented" new neural mechanism for an ancient behavior. Even if aggressive acts performed during such studies are mild and data on certain brain areas are sparse and spatially imprecise, the findings obtained so far meet evolutionary expectations. Typically, people with criminal records and those showing aggression-related psychopathologies show larger activations in the areas that are activated by the execution of aggressive acts by healthy subjects. This is also in line with rodent studies where brain areas involved in aggression were overactivated in models of abnormal aggression. In rodents, we made distinctions between hypo- and hyperarousal-associated models in this respect. The possibility of making such distinctions in humans will be discussed below.

The findings obtained in the second type of studies (called "aggression history" studies above) are quite often opposite: in broad anatomical terms, areas found activated in "manifest" studies were often found deficient in subjects with aggression histories. It is worth to stress that this is especially blatant in subjects suffering from aggression-related psychopathologies or who frequently committed violent offenses previous to the investigation. Unfortunately, comparable approaches in animal research are extremely scarce. We argued above (especially in Sect. 4.2) in favor of the idea that the seemingly contradictory findings obtained in "manifest" and "history" studies can be corroborated by assuming that the deficits seen in the latter are reflections or causes of cognitive and emotional dysfunctions that may lead to aggression indirectly. Anatomical and/or functional deficits in the prefrontal cortex and amygdala appear to result in poor moral judgment, lower ability to recognize emotions and the intentions of social partners, and increased irritability and impulsiveness. Any of these may increase the propensity to behave aggressively in critical situations, and their effects are likely nonlinearly amplified when they occur together. Even more importantly, these brain deficiency-induced psychological profiles may affect lifestyles. In contrast to experimental animals which live under fixed conditions, humans can choose where they live and what they do. The psychological deficits listed above are usually associated with risky lifestyles, where involvement in aggression is almost natural.

The considerations presented above make us to propose a dual neuronal theory of abnormal aggression in humans. A similar hypothesis was proposed in Sect. 4.2 for the prefrontal cortex; here, we extend it to the whole neurocircuitry that is involved in aggression control.

1. The *Cognitive/Emotional Mechanism.* The functional particularities of brain areas involved in decision making and emotions (e.g., the prefrontal cortex and amygdala) define the probability of being aggressive. Poor developmental conditions (e.g., unfavorable genetic and social backgrounds) adversely affect brain development, which disrupts moral judgment, the ability to recognize the intentions of social partners, and emotional responses under social pressure. These adversely affect the choice of lifestyle, which alone or in combination with brain deficits increase the likelihood that the neurobehavioral mechanism described below is activated under particular conditions. This hypothesis (illustrated in Fig. 6.2a) explains why deficits in particular brain areas increase the probability of aggressiveness and explains why deficits of variable location have similar behavioral outcomes. Judgment and emotions depend on the concerted activity of various brain mechanisms; a deficit in any of these may disrupt the functioning of the whole system.
2. *The Neurobehavioral Mechanism.* The execution of aggressive acts is a consequence of a downward cascade of neural events, in a manner very similar to that seen in rodents. The physical substrates of these events are particular regions or cell groups within the prefrontal cortex, amygdala, hypothalamus, and periaqueductal gray, which are interconnected. The triggering event originates in the prefrontal cortex (particular cell groups located within) and elicits aggressive acts by running the "command" through the chain of connections mentioned above. The acute likelihood of performing an aggressive act depends on the stimulus sensitivity of the elements of the circuitry that control aggressiveness. For example, the stimulus sensitivity of prefrontal mechanisms responsible for attack initiation defines the easiness with which decisions are transformed into behavioral acts. Similarly, the responsiveness of hypothalamic centers

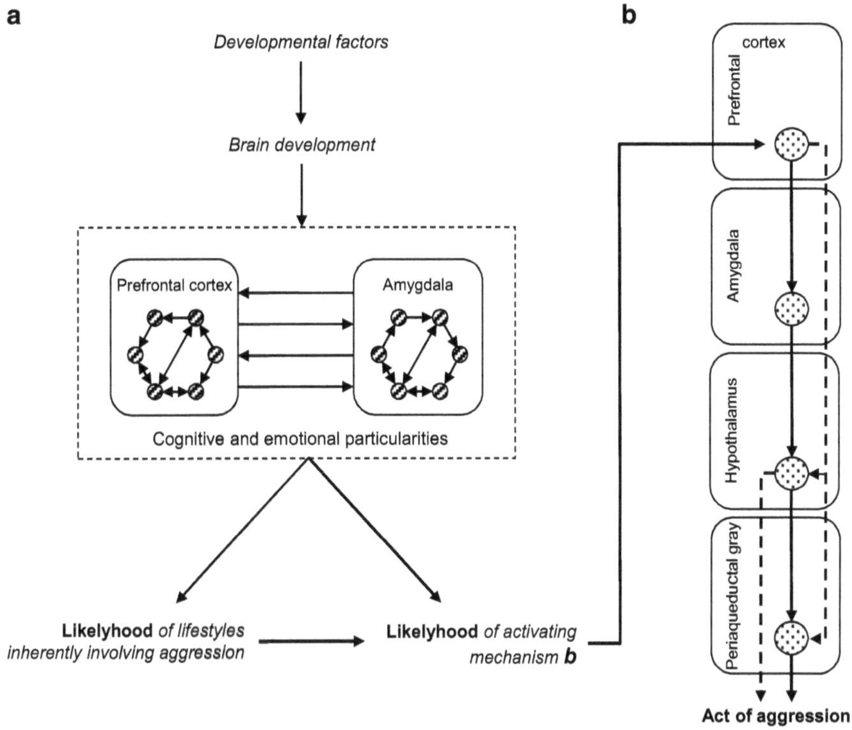

Fig. 6.2 The dual hypothesis of aggression control in humans. (**a**) Factors affecting the likelyhood of being aggressive. *Circles and arrows*, symbolic representation of brain processes involved in moral reasoning, emotion/intention recognition, and emotional control in general. (**b**) The cascade of neuronal events that lead to the execution of an aggressive act. *Circles*, subregions or cell groups within the brain areas indicated; *arrows*, the direction of information flow; *dashed arrows*, alternative or complementary routes of information flow. The *circles* symbolizing brain areas/cell groups are patterned differently in panel (**a**) and (**b**) to suggest that the neural substrates of cognitive/emotional mechanisms represented in the former are different from executory mechanisms represented in the latter. For more details see Sect. 6.2.1

determines the likelihood of transmitting the command downward to brain areas that control movements. The disruption of this information flow at any of its nodal points disrupts aggression. This explains why severe aggression is inhibited by amygdalotomy and hypothalamotomy so efficiently. Pharmacological treatments may also inhibit aggressiveness by disrupting this information flow at one or more levels. Neither of the physical substrates of the cascade includes the whole region mentioned, but subregions that cannot be identified at present. The hypothalamus may be an exception in this respect, because the so-called triangle of Sano may either be identical with or may at least include those hypothalamic subregions or cell groups that have a primary role in the elicitation of an aggressive act. This deductive hypothesis was illustrated in Fig. 6.2b.

Taken together, the dual theory of aggression control suggests that the execution of aggressive acts depends on the probability of activation of a quasi-mechanical neurobehavioral executory mechanism; the likelihood of its activation partly depends on its functional characteristics, partly on the integrity of higher psychological functions embodied in the cognitive/emotional mechanisms described above.

The dual theory is feasible only by assuming that the cognitive/emotional and neurobehavioral mechanisms are distinct. If these overlapped, their effects canceled each other out. For example, if executory and cognitive/emotional mechanisms were disrupted together because of such overlaps, increased psychological propensities to behave aggressively would lack the appropriate substrates for execution. The assumed distinctness of the two types of mechanisms would explain some of the discrepant findings presented above. For example, enhanced functioning, in particular, in prefrontal or amygdalar areas, is compatible with increased aggressiveness if increases were observed in areas that belong to the neurobehavioral executory mechanism. Moreover, the theory may also explain why changes were opposite in neighboring prefrontal and amygdalar subareas (see Figs. 4.6 and 4.11 for examples). Such opposite changes may amplify rather than cancel out each other if increased functioning was observed in areas belonging to the neurobehavioral executory mechanism, while decreases were observed in areas belonging to the cognitive/emotional mechanism.

The weak point of this theory is that it is based on assumptions and animal data rather than on evidence obtained in humans. While it is compatible with the available findings, neither the substrates of the neurobehavioral mechanism nor those of the cognitive/emotional mechanisms were differentiated so far. The advantage of the theory is that if true, it explains a great deal of the contradictory findings.

6.3.2 Types of Aggression and Brain Function

As repeatedly shown above, the findings of human brain imaging studies are highly contradictory. Contradictions may be explained by several technical reasons. For example, the technique per se is highly sophisticated, and the statistics that needs to be employed is also very complex, which together may make findings to a certain extent uncertain. Based on this, one can hypothesize that findings obtained by brain imaging techniques should be considered in general terms, and the precise comparisons presented in Sect. 4.2 are not justified (see Fig. 4.6). However, contradictions were not eliminated when the spatial resolution of analysis was decreased; findings considered at Brodmann area level (irrespective to the size and within-area location of affected structures) were no less contradictory (see Table 4.3). The number of contradictions decreases when changes are considered in broader anatomical terms (e.g., the dorsomedial prefrontal cortex), but even at this very low level of resolution, some of the contradictions remain.

The considerations outlined above either mean that brain imaging techniques should be thoroughly reevaluated regarding their validity, or one should assume that contradictions are due to another, yet poorly considered factor. Based on animal findings, we propose that this factor relates to the characteristics of subjects.

Animal studies suggest that all three behavioral characteristics, emotional background, and brain mechanisms depend on the etiological factor that led to the development of abnormal aggression. The same may be true for humans. The main problem with this approach is that etiological factors are rather complex in humans. For instance, children neglected during early phases of development are likely exposed to a variety of stressors in puberty; moreover, they bear a larger risk for drug abuse than children who were raised up under more normal conditions. As a consequence, the etiological factors of their abnormal aggressiveness may be multiple, and their relative importance is difficult to establish. Instead of etiological factors, one may group subjects based on the disorders they show and/or the types of offenses they commit. Albeit this classification may be viable, the heterogeneity of diagnostic criteria and the heterogeneity of the criminal populations studied so far may make the application of this principle rather difficult. As shown in Sect. 1.2.4 and Table 1.1, humans belonging to the very same diagnostic group may have rather different behavioral characteristics because the number of symptoms necessary for diagnosis is considerably smaller than the total number of symptoms (i.e., subjects diagnosed in the same way may have nonoverlapping symptoms). Subject populations were similarly heterogeneous regarding the type of crimes committed; the phrase "violent offense" stood for anything from wanton destructiveness to murder (see Sect. 1.2.4). While making the correspondence between brain mechanisms on one side and diagnostic criteria and violent offenses on the other side may not be an easy task retrospectively, one can reasonably assume that at least part of the contradictions noticed in the former may be eliminated by a more thorough analysis of the latter. A few recent trials where subgroups were created within diagnostic groups suggest that this approach is not only feasible but also relevant. Studies making broader distinctions, e.g., those differentiating reactive from proactive aggression, may also be highly relevant in this respect.

Taken together, we propose here that similar to animals, "abnormal aggression" has no unitary underlying mechanism in humans. The main arguments supporting this assumption are the diversity of brain findings and the similar diversity of subjects. Finding the relationship between particular constellations of cognitive, emotional, and behavioral features on one side and a similarly intricate constellation of neural alterations on the other side is not easy and remains a task for the future.

6.4 Hormones

Instead of influencing behavior directly, hormones define the properties of neurons. Their effects are primarily relevant for brain function and only indirectly to behavior. Nevertheless, hormonal effects are quite important in aggression research, where brain mechanisms are incompletely understood.

Data in animals and humans seem to converge in the meaning that each particular form of abnormal aggression is associated with particular changes in brain function. While the understanding of these may require considerable effort, the elucidation of hormonal backgrounds is relatively easy. The production patterns of testosterone and glucocorticoids and their relationship received much attention in recent years and rightfully so: while one can exclude that any of these or other hormones trigger the execution of aggressive acts directly, they have an important impact on brain development and function.

Making the correspondence between hormone production patterns and behavior may be important in two ways. On one hand, the findings of this line of research may be used as classification principles for differentiating types of abnormal aggression; as such findings with hormones—together with those that elucidate other aspects of emotionality—may be used to guide the differentiation of aggression types for brain studies. On the other hand, endocrine functions are highly dependent on environmental and genetic backgrounds and may be among those factors that make the link between these and alterations in brain function. Consequently, changes in hormone production—together with epigenetic mechanisms that are also under hormonal control—may be explanatory in terms of the developmental routes that lead to particular forms of aggression.

Postscript

The understanding of brain mechanisms underlying aggression developed tremendously since it was discovered that the electrical stimulation of particular brain regions induces goal-directed aggression in animals. The areas involved and their connections were identified to a large extent over the last century. One line of reasoning suggested that brain mechanisms revealed in animals are working in humans and attributed socially dangerous forms of aggression to alterations in the neural mechanisms that were discovered in animals. This view was supported by the consequences of certain surgical interventions and pharmacological treatments. However, surgical interventions were abandoned for ethical reasons and for the perceived unspecificity of their effects. Pharmacological treatments on their turn were not always effective; moreover, their efficacy was to large extent unpredictable. These findings lent support to a somewhat older theory suggesting that human aggression is a result of misconceptions on self, others, and society as a whole and to deficits in emotional control. While the role of biological factors was not completely negated, a great deal of research into human aggression considered them of marginal importance. The two lines of reasoning remained rather detached during the last century, and as a result, two mainstreams developed in aggression research: one focused on biological mechanisms in animals, while the other gradually deciphered the social and psychological factors underlying human aggression.

One of the most important contributions of brain imaging techniques to the understanding of the human mind was the discovery that brain functions, cognition and emotions are tightly linked. As a consequence, psychological and neural mechanisms of aggression control cannot be separated any more. In parallel, several abnormal aggression models were developed in animals. These showed that psychological and social factors leading to abnormal aggression in humans can be modeled in rodents, and these lead to abnormal manifestations of aggressiveness; moreover, many aggression-independent psychological deficits seen in abnormally aggressive people are also reproduced in rodent models of abnormal aggression. These two lines of evidence show that animal and human aggression are not as detached as previously thought; similar developmental factors induce comparable changes in cognition, emotion and behavior, and the evidence

J. Haller, *Neurobiological Bases of Abnormal Aggression and Violent Behaviour*, DOI 10.1007/978-3-7091-1268-7, © Springer-Verlag Wien 2014

accumulated so far suggests that changes in brain mechanisms are also comparable. To date, no reasons seem to remain in favor of the idea that brain mechanisms should be studied in animals and social/psychological mechanisms should be studied in humans.

At present, neither animal nor human research seems to be at its apogee regarding the elucidation of cognitive/emotional and biological bases of aggression in general and of abnormal aggression in particular. This is reflected by both the large body of contradictory evidence and the scarcity of knowledge in many subfields. Understanding seems to be hampered by both technical difficulties and deficient conceptual approaches. Techniques rapidly improve; the difficulties faced today are likely overcome in the near future. Concepts also evolve, and simplistic approaches gradually seem to give way to more workable ones that take into account the complexities of both brain function and behavior. From this perspective, aggression research seems to step into a new era, where shortcomings of the past pave the ways of the better approaches of tomorrow.

Partly supported by the Hungarian National Science and Research Fund (OTKA). Contract grant number: 82069.

References

Abler B, Walter H, Erk S (2005) Neural correlates of frustration. Neuroreport 16:669–672

Adamantidis AR, Zhang F, Aravanis AM, Deisseroth K, de Lecea L (2007) Neural substrates of awakening probed with optogenetic control of hypocretin neurons. Nature 450:420–424

Adams DB (1968) Cells related to fighting behavior recorded from midbrain central gray neuropil of cat. Science 159:894–896

Adams DB (2006) Brain mechanisms of aggressive behavior: an updated review. Neurosci Biobehav Rev 30:304–318

Adams DB, Boudreau W, Cowan CW, Kokonowski C, Oberteuffer K, Yohay K (1993) Offense produced by chemical stimulation of the anterior hypothalamus of the rat. Physiol Behav 53:1127–1132

Adriani W, Leo D, Greco D, Rea M, di Porzio U, Laviola G, Perrone-Capano C (2006) Methylphenidate administration to adolescent rats determines plastic changes on reward-related behavior and striatal gene expression. Neuropsychopharmacology 31:1946–1956

Ago Y, Araki R, Tanaka T, Sasaga A, Nishiyama S, Takuma K, Matsuda T (2013) Role of social encounter-induced activation of prefrontal serotonergic systems in the abnormal behaviors of isolation-reared mice. Neuropsychopharmacology 38(8):1535–1547

Akbari EM, Budin R, Parada M, Fleming AS (2008) The effects of early isolation on sexual behavior and c-fos expression in naïve male Long-Evans rats. Dev Psychobiol 50:298–306

Albers HE, Dean A, Karom MC, Smith D, Huhman KL (2006) Role of V1a vasopressin receptors in the control of aggression in Syrian hamsters. Brain Res 1073–1074:425–430

Albert DJ, Walsh ML (1984) Neural systems and the inhibitory modulation of agonistic behavior: a comparison of mammalian species. Neurosci Biobehav Rev 8:5–24

Albert DJ, Walsh ML, Zalys C, Dyson E (1986) Defensive aggression toward an experimenter: no differences between males and females following septal, medial accumbens, or medial hypothalamic lesions in rats. Physiol Behav 38:11–14

Albert DJ, Walsh ML, Jonik RH (1993) Aggression in humans: what is its biological foundation? Neurosci Biobehav Rev 17:405–425

Alden LE, Taylor CT (2004) Interpersonal processes in social phobia. Clin Psychol Rev 24:857–882

Allen A, Links PS (2012) Aggression in borderline personality disorder: evidence for increased risk and clinical predictors. Curr Psychiatry Rep 14:62–69

Almeida M, Lee R, Coccaro EF (2010) Cortisol responses to ipsapirone challenge correlate with aggression, while basal cortisol levels correlate with impulsivity, in personality disorder and healthy volunteer subjects. J Psychiatr Res 44(14):874–880

Aluja A, García LF (2007) Role of sex hormone-binding globulin in the relationship between sex hormones and antisocial and aggressive personality in inmates. Psychiatry Res 152:189–196

J. Haller, *Neurobiological Bases of Abnormal Aggression and Violent Behaviour*,
DOI 10.1007/978-3-7091-1268-7, © Springer-Verlag Wien 2014

Ambar G, Chiavegatto S (2009) Anabolic-androgenic steroid treatment induces behavioral disinhibition and downregulation of serotonin receptor messenger RNA in the prefrontal cortex and amygdala of male mice. Genes Brain Behav 8:161–173

Amen DG, Stubblefield M, Carmicheal B, Thisted R (1996) Brain SPECT findings and aggressiveness. Ann Clin Psychiatry 8:129–137

American Psychiatric Association (2000) Diagnostic and statistical manual of mental disorders. Text rev., 4th edn. APA, Washington, DC

Amore M, Menchetti M, Tonti C, Scarlatti F, Lundgren E, Esposito W, Berardi D (2008) Predictors of violent behavior among acute psychiatric patients: clinical study. Psychiatry Clin Neurosci 62:247–255

Amstislavskaya TG, Bulygina VV, Tikhonova MA, Maslova LN (2013) Social isolation during peri-adolescence or adulthood: effects on sexual motivation, testosterone and corticosterone response under conditions of sexual arousal in male rats. Chin J Physiol 56:36–43

Amunts K, Kedo O, Kindler M, Pieperhoff P, Mohlberg H, Shah NJ, Habel U, Schneider F, Zilles K (2005) Cytoarchitectonic mapping of the human amygdala, hippocampal region and entorhinal cortex: intersubject variability and probability maps. Anat Embryol 210:343–352

Anderson DJ (2012) Optogenetics, sex, and violence in the brain: implications for psychiatry. Biol Psychiatry 71:1081–1089

Anderson NE, Kiehl KA (2012) The psychopath magnetized: insights from brain imaging. Trends Cogn Sci 16:52–60

Anderson SW, Bechara A, Damasio H, Tranel D, Damasio AR (1999) Impairment of social and moral behavior related to early damage in human prefrontal cortex. Nat Neurosci 2:1032–1037

Anderson CA, Shibuya A, Ihori N, Swing EL, Bushman BJ, Sakamoto A, Rothstein HR, Saleem M (2010) Violent video game effects on aggression, empathy, and prosocial behavior in eastern and western countries: a meta-analytic review. Psychol Bull 136:151–173

Ansah TA, Wade LH, Shockley DC (1996) Changes in locomotor activity, core temperature, and heart rate in response to repeated cocaine administration. Physiol Behav 60:1261–1267

Anschel S, Alexander M, Perachio AA (1982) Multiple connections of medial hypothalamic neurons in the rat. Exp Brain Res 46:383–392

Antonucci AS, Gansler DA, Tan S, Bhadelia R, Patz S, Fulwiler C (2006) Orbitofrontal correlates of aggression and impulsivity in psychiatric patients. Psychiatry Res 147:213–220

Aravanis AM, Wang LP, Zhang F, Meltzer LA, Mogri MZ, Schneider MB, Deisseroth K (2007) An optical neural interface: in vivo control of rodent motor cortex with integrated fiberoptic and optogenetic technology. J Neural Eng 4:S143–S156

Archer J (1991) The influence of testosterone on human aggression. Br J Psychol 82:1–28

Archer J (2006) Testosterone and human aggression: an evaluation of the challenge hypothesis. Neurosci Biobehav Rev 30:319–345

Archer J (2009) The nature of human aggression. Int J Law Psychiatry 32:202–208

Archer J, Benson D (2008) Physical aggression as a function of perceived fighting ability and provocation: an experimental investigation. Aggress Behav 34:9–24

Archer J, Ireland JL, Power CL (2007) Differences between bullies and victims, and men and women, on aggression-related variables among prisoners. Br J Soc Psychol 46:299–322

Attar BK, Guerra NG, Tolan PH (1994) Neighborhood disadvantage, stressful life events, and adjustment in urban elementary-school children. J Clin Child Psychol 23:391–400

Auyeung B, Baron-Cohen S, Ashwin E, Knickmeyer R, Taylor K, Hackett G, Hines M (2009) Fetal testosterone predicts sexually differentiated childhood behavior in girls and in boys. Psychol Sci 20:144–148

Azar R, Zoccolillo M, Paquette D, Quiros E, Baltzer F, Tremblay RE (2004) Cortisol levels and conduct disorder in adolescent mothers. J Am Acad Child Adolesc Psychiatry 43:461–468

Bacskai E, Czobor P, Gerevich J (2011) Gender differences in trait aggression in young adults with drug and alcohol dependence compared to the general population. Prog Neuropsychopharmacol Biol Psychiatry 35:1333–1340

Badre D, D'Esposito M (2009) Is the rostro-caudal axis of the frontal lobe hierarchical? Nat Rev Neurosci 10:659–669
Bailey AA, Hurd PL (2005) Finger length ratio (2D:4D) correlates with physical aggression in men but not in women. Biol Psychol 68:215–222
Balasubramaniam V, Kanaka TS (1975) Amygdalotomy and hypothalamotomy – a comparative study. Confin Neurol 37:195–201
Balasubramanian K, Pereira BM, Govindarajulu P (1983) Effect of adrenalectomy and corticosterone replacement on epididymal carbohydrate metabolism-studies on mature male rats. Andrologia 15:71–77
Baldwin JJ, Randolph DL (1982) The effects of a provocation on aggression for three types of alcohol users. J Clin Psychol 38:439–444
Balthazart J, Cornil CA, Charlier TD, Taziaux M, Ball GF (2009) Estradiol, a key endocrine signal in the sexual differentiation and activation of reproductive behavior in quail. J Exp Zool A Ecol Genet Physiol 311:323–345
Bandler RJ Jr (1970) Cholinergic synapses in the lateral hypothalamus for the control of predatory aggression in the rat. Brain Res 20:409–424
Bandler RJ (1977) Predatory behavior in the cat elicited by lower brain stem and hypothalamic stimulation: a comparison. Brain Behav Evol 14:440–460
Bandler R (1982) Induction of 'rage' following microinjections of glutamate into midbrain but not hypothalamus of cats. Neurosci Lett 30:183–188
Bandler R, Depaulis A (1991) Midbrain periaqueductal gray control of defensive behavior in the cat and the rat. In: Depaulis A, Bandler R (eds) The midbrain periaqueductal gray matter: functional, anatomical and neurochemical organization. Plenum, New York, NY
Bandler RJ Jr, Flynn JP (1974) Neural pathways from thalamus associated with regulation of aggressive behavior. Science 183:96–99
Bandler R, Shipley MT (1994) Columnar organization in the midbrain periaqueductal gray: modules for emotional expression? Trends Neurosci 17:379–389
Bandler R, Keay KA, Floyd N, Price J (2000) Central circuits mediating patterned autonomic activity during active vs. passive emotional coping. Brain Res Bull 53:95–104
Banks T, Dabbs JM Jr (1996) Salivary testosterone and cortisol in a delinquent and violent urban subculture. J Soc Psychol 136:49–56
Barkataki I, Kumari V, Das M, Taylor P, Sharma T (2006) Volumetric structural brain abnormalities in men with schizophrenia or antisocial personality disorder. Behav Brain Res 169:239–247
Baron RA (1977) Human aggression. Plenum, New York, NY
Barrash J, Tranel D, Anderson SW (2000) Acquired personality disturbances associated with bilateral damage to the ventromedial prefrontal region. Dev Neuropsychol 18:355–381
Barrash J, Asp E, Markon K, Manzel K, Anderson SW, Tranel D (2011) Dimensions of personality disturbance after focal brain damage: investigation with the Iowa Scales of Personality Change. J Clin Exp Neuropsychol 33:833–852
Bauman MD, Toscano JE, Mason WA, Lavenex P, Amaral DG (2006) The expression of social dominance following neonatal lesions of the amygdala or hippocampus in rhesus monkeys (Macaca mulatta). Behav Neurosci 120:749–760
Beach SR, Brody GH, Todorov AA, Gunter TD, Philibert RA (2011) Methylation at 5HTT mediates the impact of child sex abuse on women's antisocial behavior: an examination of the Iowa adoptee sample. Psychosom Med 73:83–87
Bear DM, Fulop M (1987) The neurology of emotion. Arch Neurol Psychiatry 42:979–1000
Beauchaine TP, Neuhaus E, Zalewski M, Crowell SE, Potapova N (2011) The effects of allostatic load on neural systems subserving motivation, mood regulation, and social affiliation. Dev Psychopathol 23:975–999
Beaver JD, Lawrence AD, Passamonti L, Calder AJ (2008) Appetitive motivation predicts the neural response to facial signals of aggression. J Neurosci 28:2719–2725

Bedrosian TA, Fonken LK, Demas GE, Nelson RJ (2012) Photoperiod-dependent effects of neuronal nitric oxide synthase inhibition on aggression in Siberian hamsters. Horm Behav 61:176–180

Beeman EA (1947) The effect of male hormone on aggressive behavior in mice. Physiol Zool 20:373–405

Behbehani MM (1995) Functional characteristics of the midbrain periaqueductal gray. Prog Neurobiol 46(6):575–605

Beiderbeck DI, Reber SO, Havasi A, Bredewold R, Veenema AH, Neumann ID (2012) High and abnormal forms of aggression in rats with extremes in trait anxiety – involvement of the dopamine system in the nucleus accumbens. Psychoneuroendocrinology 37:1969–1980

Beitchman JH, Zai CC, Muir K, Berall L, Nowrouzi B, Choi E, Kennedy JL (2012) Childhood aggression, callous-unemotional traits and oxytocin genes. Eur Child Adolesc Psychiatry. PMID: 22294460 [Epub ahead of print]

Bejjani BP, Houeto JL, Hariz M, Yelnik J, Mesnage V, Bonnet AM, Pidoux B, Dormont D, Cornu P, Agid Y (2002) Aggressive behavior induced by intraoperative stimulation in the triangle of Sano. Neurology 59:1425–1427

Bellani M, Baiano M, Brambilla P (2011) Brain anatomy of major depression II. Focus on amygdala. Epidemiol Psychiatr Sci 20:33–36

Beltt BM, Keesey RE (1975) Hypothalamic map of stimulation current thresholds for inhibition of feeding in rats. Am J Physiol 229:1124–1133

Benderlioglu Z, Nelson RJ (2004) Digit length ratios predict reactive aggression in women, but not in men. Horm Behav 46:558–564

Benderlioglu Z, Sciulli PW, Nelson RJ (2004) Fluctuating asymmetry predicts human reactive aggression. Am J Hum Biol 16:458–469

Benis AM, Hobgood DK (2011) Dopamine receptor DRD3 codes for trait aggression as Mendelian recessive. Med Hypotheses 77:1108–1110

Bennazzous A, Hallett M (2000) Mechanism of action of deep brain stimulation. Neurology 55: S13–S16

Bennett RM, Buss AH, Carpenter JA (1969) Alcohol and human physical aggression. Quart J Stud Alcohol 30:870–876

Benus RF, Bohus B, Koolhaas JM, van Oortmerssen GA (1991) Behavioural differences between artificially selected aggressive and non-aggressive mice: response to apomorphine. Behav Brain Res 43:203–208

Berdahl CH (2010) A neural network model of borderline personality disorder. Neural Netw 23:177–188

Berenbaum SA, Duck SC, Bryk K (2000) Behavioral effects of prenatal versus postnatal androgen excess in children with 21-hydroxylase-deficient congenital adrenal hyperplasia. J Clin Endocrinol Metab 85:727–733

Bergeon Burns CM, Rosvall KA, Ketterson ED (2013) Neural steroid sensitivity and aggression: comparing individuals of two songbird subspecies. J Evol Biol 26:820–831

Bergman B, Brismar B (1994) Hormone levels and personality traits in abusive and suicidal male alcoholics. Alcohol Clin Exp Res 18:311–316

Bergvall AH, Wessely H, Forsman A, Hansen S (2001) A deficit in attentional set-shifting of violent offenders. Psychol Med 31:1095–1105

Berkowitz L (1993) The problem of aggression. In: Berkowitz L (ed) Aggression: its causes, consequences, and control, McGraw-Hill series in social psychology. McGraw-Hill, New York, NY, pp 1–23

Berkowitz L (1994) Aggression: definition and perspectives. In: Baron RA, Richardson DR (eds) Perspectives in social psychology: human aggression, 2nd edn. Plenum, New York, NY, pp 1–38

Berlin HA, Rolls ET, Kischka U (2004) Impulsivity, time perception, emotion and reinforcement sensitivity in patients with orbitofrontal cortex lesions. Brain 127:1108–1126

Berlin HA, Rolls ET, Iversen SD (2005) Borderline personality disorder, impulsivity, and the orbitofrontal cortex. Am J Psychiatry 162:2360–2373
Berman ME, Tracy JI, Coccaro EF (1997) The serotonin hypothesis of aggression revisited. Clin Psychol Rev 17:651–665
Berridge KC, Valenstein ES (1991) What psychological process mediates feeding evoked by electrical stimulation of the lateral hypothalamus? Behav Neurosci 105:3–14
Bertsch K, Grothe M, Prehn K, Vohs K, Berger C, Hauenstein K, Keiper P, Domes G, Teipel S, Herpertz SC (2013) Brain volumes differ between diagnostic groups of violent criminal offenders. Eur Arch Psychiatry Clin Neurosci 263:593–606
Best M, Williams JM, Coccaro EF (2002) Evidence for a dysfunctional prefrontal circuit in patients with an impulsive aggressive disorder. Proc Natl Acad Sci U S A 99:8448–8453
Bethea CL, Reddy AP, Robertson N, Coleman K (2013) Effects of aromatase inhibition and androgen activity on serotonin and behavior in male macaques. Behav Neurosci 127:400–414
Bhatt S, Siegel A (2006) Potentiating role of interleukin 2 (IL-2) receptors in the midbrain periaqueductal gray (PAG) upon defensive rage behavior in the cat: role of neurokinin NK (1) receptors. Behav Brain Res 167:251–260
Bhatt S, Zalcman S, Hassanain M, Siegel A (2005) Cytokine modulation of defensive rage behavior in the cat: role of GABAA and interleukin-2 receptors in the medial hypothalamus. Neuroscience 133:17–28
Bhatt S, Bhatt R, Zalcman SS, Siegel A (2008) Role of IL-1 beta and 5-HT2 receptors in midbrain periaqueductal gray (PAG) in potentiating defensive rage behavior in cat. Brain Behav Immun 22:224–233
Bibancos T, Jardim DL, Aneas I, Chiavegatto S (2007) Social isolation and expression of serotonergic neurotransmission-related genes in several brain areas of male mice. Genes Brain Behav 6:529–539
Bie B, Brown DL, Naguib M (2011) Synaptic plasticity and pain aversion. Eur J Pharmacol 667:26–31
Birbaumer N, Veit R, Lotze M, Erb M, Hermann C, Grodd W, Flor H (2005) Deficient fear conditioning in psychopathy: a functional magnetic resonance imaging study. Arch Gen Psychiatry 62:799–805
Bittencourt AS, Nakamura-Palacios EM, Mauad H, Tufik S, Schenberg LC (2005) Organization of electrically and chemically evoked defensive behaviors within the deeper collicular layers as compared to the periaqueductal gray matter of the rat. Neuroscience 133:873–892
Bjorkqvist K, Niemela P (1992) New trends in the study of female aggression. In: Bjoerkqvist K, Niemelae P (eds) Of mice and women: aspects of female aggression. Academic, San Diego, CA, pp 3–16
Bjorkqvist K, Oesterman K, Kaukiainen A (1992) The development of direct and indirect aggressive strategies in males and females. In: Bjoerkqvist K, Niemelae P (eds) Of mice and women: aspects of female aggression. Academic, San Diego, CA, pp 51–64
Björkqvist K, Lindström M, Pehrsson M (2000) Attribution of aggression to acts: a four-factor model. Psychol Rep 87:525–530
Blair RJ (2001) Neurocognitive models of aggression, the antisocial personality disorders, and psychopathy. J Neurol Neurosurg Psychiatry 71:727–731
Blair RJ (2007) Dysfunctions of medial and lateral orbitofrontal cortex in psychopathy. Ann NY Acad Sci 1121:461–479
Blair RJ (2010a) Neuroimaging of psychopathy and antisocial behavior: a targeted review. Curr Psychiatry Rep 12:76–82
Blair RJ (2010b) Psychopathy, frustration, and reactive aggression: the role of ventromedial prefrontal cortex. Br J Psychol 101:383–399
Blair RJR (2012) Considering anger from a cognitive neuroscience perspective. Wiley Interdiscip Rev Cogn Sci 3:65–74
Blair RJ, Cipolotti L (2000) Impaired social response reversal. A case of 'acquired sociopathy'. Brain 123:1122–1141

Blair RJ, Morris JS, Frith CD, Perrett DI, Dolan RJ (1999) Dissociable neural responses to facial expressions of sadness and anger. Brain 122:883–893

Blair C, Granger D, Razza RP (2005) Cortisol reactivity is positively related to executive function in preschool children attending Head Start. Child Dev 76:554–567

Blanchard RJ, Blanchard DC (1989) Attack and defense in rodents as ethoexperimental models for the study of emotion. Prog Neuropsychopharmacol Biol Psychiatry 13(Suppl):S3–S14

Blanchard DC, Takahashi SN (1988) No change in intermale aggression after amygdala lesions which reduce freezing. Physiol Behav 42:613–616

Blanchard RJ, Flannelly KJ, Blanchard DC (1988) Life-span studies of dominance and aggression in established colonies of laboratory rats. Physiol Behav 43:1–7

Blanchard RJ, Blanchard DC, Rodgers J, Weiss SM (1990) The characterization and modelling of antipredator defensive behavior. Neurosci Biobehav Rev 14:463–472

Blanchard DC, Spencer RL, Weiss SM, Blanchard RJ, McEwen B, Sakai RR (1995) Visible burrow system as a model of chronic social stress: behavioral and neuroendocrine correlates. Psychoneuroendocrinology 20:117–134

Blanchard RJ, Griebel G, Henrie JA, Blanchard DC (1997) Differentiation of anxiolytic and panicolytic drugs by effects on rat and mouse defense test batteries. Neurosci Biobehav Rev 21:783–789

Blanchard RJ, Wall PM, Blanchard DC (2003) Problems in the study of rodent aggression. Horm Behav 44:161–170

Bliss-Moreau E, Moadab G, Bauman MD, Amaral DG (2013) The impact of early amygdala damage on juvenile rhesus macaque social behavior. J Cogn Neurosci 25:2124–2140

Blomen VA, Boonstra J (2011) Stable transmission of reversible modifications: maintenance of epigenetic information through the cell cycle. Cell Mol Life Sci 68:27–44

Blumer D, Benson DF (1975) Personality changes with frontal and temporal lobe lesions. In: Benson DF, Blumer D (eds) Psychiatric aspects of neurological disease. Grune & Stratton, New York, NY, pp 151–169

Boccardi M, Frisoni GB, Hare RD, Cavedo E, Najt P, Pievani M, Rasser PE, Laakso MP, Aronen HJ, Repo-Tiihonen E, Vaurio O, Thompson PM, Tiihonen J (2011) Cortex and amygdala morphology in psychopathy. Psychiatry Res 193:85–92

Boes AD, Tranel D, Anderson SW, Nopoulos P (2008) Right anterior cingulate: a neuroanatomical correlate of aggression and defiance in boys. Behav Neurosci 122:677–684

Böhnke R, Bertsch K, Kruk MR, Naumann E (2010a) The relationship between basal and acute HPA axis activity and aggressive behavior in adults. J Neural Transm 117:629–637

Böhnke R, Bertsch K, Kruk MR, Richter S, Naumann E (2010b) Exogenous cortisol enhances aggressive behavior in females, but not in males. Psychoneuroendocrinology 35:1034–1044

Boncoraglio G, Groothuis TG (2013) Effect of rapid modulation of circulating plasma testosterone concentration on begging, aggressive behavior and competition for food in black-headed gull (Larus ridibundus) chicks. Horm Behav 64(3):487–493

Bonnet U (2003) Moclobemide: therapeutic use and clinical studies. CNS Drug Rev 9:97–140

Booij L, Tremblay RE, Leyton M, Séguin JR, Vitaro F, Gravel P, Perreau-Linck E, Lévesque ML, Durand F, Diksic M, Turecki G, Benkelfat C (2010) Brain serotonin synthesis in adult males characterized by physical aggression during childhood: a 21-year longitudinal study. PLoS One 5:e11255

Book AS, Starzyk KB, Quinsey VL (2001) The relationship between testosterone and aggression: a meta-analysis. Aggress Viol Behav 6:579–599

Booth A, Osgood D (1993) The influence of testosterone on deviance in adulthood. Criminology 31:93–117

Booth A, Shelley G, Mazur A, Tharp G, Kittok R (1989) Testosterone, and winning and losing in human competition. Horm Behav 23(4):556–571

Bortolato M, Chen K, Godar SC, Chen G, Wu W, Rebrin I, Farrell MR, Scott AL, Wellman CL, Shih JC (2011) Social deficits and perseverative behaviors, but not overt aggression, in MAO-A hypomorphic mice. Neuropsychopharmacology 36:2674–2688

Bortolato M, Godar SC, Melis M, Soggiu A, Roncada P, Casu A, Flore G, Chen K, Frau R, Urbani A, Castelli MP, Devoto P, Shih JC (2012) NMDARs mediate the role of monoamine oxidase A in pathological aggression. J Neurosci 32:8574–8582

Bountra C, Oppermann U, Heightman TD (2011) Animal models of epigenetic regulation in neuropsychiatric disorders. Curr Top Behav Neurosci 7:281–322

Boyden ES, Zhang F, Bamberg E, Nagel G, Deisseroth K (2005) Millisecond-timescale, genetically targeted optical control of neural activity. Nat Neurosci 8:1263–1268

Brain PF, Nowell NW, Wouters A (1971) Some relationships between adrenal function and isolation induced intermale aggression in albino mice. Physiol Behav 6:27–29

Brambilla F, Bellodi L, Arancio C, Limonta D, Ferrari E, Solerte B (2001) Neurotransmitter and hormonal background of hostility in anorexia nervosa. Neuropsychobiology 43:225–232

Brambilla P, Soloff PH, Sala M, Nicoletti MA, Keshavan MS, Soares JC (2004) Anatomical MRI study of borderline personality disorder patients. Psychiatry Res 131:125–133

Breuillaud L, Rossetti C, Meylan EM, Mérinat C, Halfon O, Magistretti PJ, Cardinaux JR (2012) Deletion of CREB-regulated transcription coactivator 1 induces pathological aggression, depression-related behaviors, and neuroplasticity genes dysregulation in mice. Biol Psychiatry 72:528–536

Brewer-Smyth K, Burgess AW, Shults J (2004) Physical and sexual abuse, salivary cortisol, and neurologic correlates of violent criminal behavior in female prison inmates. Biol Psychiatry 55:21–31

Bronson FH, Desjardins C (1968) Aggression in adult mice: modification by neonatal injections of gonadal hormones. Science 161:705–706

Bronson FH, Desjardins C (1969) Aggressive behavior and seminal vesicle function in mice: differential sensitivity to androgen given neonatally. Endocrinology 85:971–974

Bronson FH, Eleftheriou BE (1965) Adrenal response to fighting in mice: separation of physical and psychological causes. Science 147:627–628

Brooks JH, Reddon JR (1996) Serum testosterone in violent and nonviolent young offenders. J Clin Psychol 52:475–483

Brower MC, Price BH (2001) Neuropsychiatry of frontal lobe dysfunction in violent and criminal behaviour: a critical review. J Neurol Neurosurg Psychiatry 71:720–726

Brown GL, Goodwin FK, Ballenger JC, Goyer PF (1979) Major aggression in human correlates with cerebrospinal fluid amine metabolites. Psychiatry Res 1:131–139

Brunner HG, Nelen M, Breakefield XO, Ropers HH, van Oost BA (1993) Abnormal behavior associated with a point mutation in the structural gene for monoamine oxidase A. Science 262:578–580

Brutus M, Siegel A (1989) Effects of the opiate antagonist naloxone upon hypothalamically elicited affective defense behavior in the cat. Behav Brain Res 33:23–32

Brutus M, Watson RE Jr, Shaikh MB, Siegel HE, Weiner S, Siegel A (1984) A [14C]2-deoxyglucose analysis of the functional neural pathways of the limbic forebrain in the rat. IV. A pathway from the prefrontal cortical-medial thalamic system to the hypothalamus. Brain Res 310:279–293

Brutus M, Shaikh MB, Edinger H, Siegel A (1986) Effects of experimental temporal lobe seizures upon hypothalamically elicited aggressive behavior in the cat. Brain Res 366:53–63

Bunck M, Czibere L, Horvath C, Graf C, Frank E, Kessler MS, Murgatroyd C, Müller-Myhsok B, Gonik M, Weber P, Pütz B, Muigg P, Panhuysen M, Singewald N, Bettecken T, Deussing JM, Holsboer F, Spengler D, Landgraf R (2009) A hypomorphic vasopressin allele prevents anxiety-related behavior. PLoS One 4:e5129

Bushman BJ, Anderson CA (2001) Is it time to pull the plug on hostile versus instrumental aggression dichotomy? Psychol Rev 108:273–279

Buss AH (1961) The psychology of aggression. Wiley, New York, NY

Buss AH, Durkee A (1957) An inventory for assessing different kinds of hostility. J Cons Psychol 21:343–349

Buss AH, Perry M (1992) The aggression questionnaire. J Pers Soc Psychol 63:452–459

Butovskaya ML, Vasilyev VA, Lazebny OE, Burkova VN, Kulikov AM, Mabulla A, Shibalev DV, Ryskov AP (2012) Aggression, digit ratio, and variation in the androgen receptor, serotonin transporter, and dopamine D4 receptor genes in African foragers: the Hadza. Behav Genet 42:647–662

Butovskaya M, Fedenok J, Burkova V, Manning J (2013) Sex differences in 2D:4D and aggression in children and adolescents from five regions of Russia. Am J Phys Anthropol 152(1):130–139

Butterfield PA, Crook JH (1968) The annual cycle of nest building and agonistic behaviour in captive Quelea quelea with reference to endocrine factors. Anim Behav 16:308–317

Buydens-Branchey L, Branchey M (2004) Cocaine addicts with conduct disorder are typified by decreased cortisol responsivity and high plasma levels of DHEA-S. Neuropsychobiology 50:161–166

Bzdok D, Laird AR, Zilles K, Fox PT, Eickhoff SB (2013) An investigation of the structural, connectional, and functional subspecialization in the human amygdala. Hum Brain Mapp 34:3247–3266

Caffrey MK, Nephew BC, Febo M (2010) Central vasopressin V1a receptors modulate neural processing in mothers facing intruder threat to pups. Neuropharmacology 58:107–116

Caillé I, Dumartin B, Bloch B (1996) Ultrastructural localization of D1 dopamine receptor immunoreactivity in rat striatonigral neurons and its relation with dopaminergic innervation. Brain Res 730:17–31

Caldji C, Diorio J, Meaney MJ (2000) Variations in maternal care in infancy regulate the development of stress reactivity. Biol Psychiatry 48:1164–1174

Caldwell GS, Glickman SE, Smith ER (1984) Seasonal aggression independent of seasonal testosterone in wood rats. Proc Natl Acad Sci 81:5255–5257

Calhoon GG, O'Donnell P (2013) Closing the gate in the limbic striatum: prefrontal suppression of hippocampal and thalamic inputs. Neuron 78:181–190

Cambon K, Dos-Santos Coura R, Groc L, Carbon A, Weissmann D, Changeux JP, Pujol JF, Granon S (2010) Aggressive behavior during social interaction in mice is controlled by the modulation of tyrosine hydroxylase expression in the prefrontal cortex. Neuroscience 171:840–851

Canoine V, Gwinner E (2002) Seasonal differences in the hormonal control of territorial aggression in free-living European stonechats. Horm Behav 41:1–8

Canoine V, Fusani L, Schlinger B, Hau M (2007) Low sex steroids, high steroid receptors: increasing the sensitivity of the nonreproductive brain. Dev Neurobiol 67:57–67

Canteras NS, Mota-Ortiz SR, Motta SC (2012) What ethologically based models have taught us about the neural systems underlying fear and anxiety. Braz J Med Biol Res 45:321–327

Capaldi DM, Rothbart MK (1992) Development and validation of an early adolescent temperament measure. J Early Adolesc 12:153–173

Caramaschi D, de Boer SF, Koolhaas JM (2007) Differential role of the 5-HT1A receptor in aggressive and non-aggressivemice: an across-strain comparison. Physiol Behav 90:590–601

Caramaschi D, de Boer SF, de Vries H, Koolhaas JM (2008a) Development of violence in mice through repeated victory along with changes in prefrontal cortex neurochemistry. Behav Brain Res 189:263–272

Caramaschi D, de Boer SF, Koolhaas JM (2008b) Is hyper-aggressiveness associated with physiological hypoarousal? A comparative study on mouse lines selected for high and low aggressiveness. Physiol Behav 95:591–598

Carrasco JL, Tajima-Pozo K, Díaz-Marsá M, Casado A, López-Ibor JJ, Arrazola J, Yus M (2012) Microstructural white matter damage at orbitofrontal areas in borderline personality disorder. J Affect Disord 139:149–153

Carré JM, Gilchrist JD, Morrissey MD, McCormick CM (2010) Motivational and situational factors and the relationship between testosterone dynamics and human aggression during competition. Biol Psychol 84:346–353

Carré JM, Murphy KR, Hariri AR (2013) What lies beneath the face of aggression? Soc Cogn Affect Neurosci 8:224–229

Carrillo M, Ricci LA, Melloni RH Jr (2009) Adolescent anabolic androgenic steroids reorganize the glutamatergic neural circuitry in the hypothalamus. Brain Res 1249:118–127

Carrillo M, Ricci LA, Melloni RH (2011) Developmental and withdrawal effects of adolescent AAS exposure on the glutamatergic system in hamsters. Behav Neurosci 125:452–464

Carrive P, Morgan MM (2012) Periaqueductal gray. In: Paxinos G, Mai JK (eds) The human nervous system. Academic, London

Carvalho FS, Beblo T, Schlosser N, Terfehr K, Otte C, Löwe B, Wolf OT, Spitzer C, Driessen M, Wingenfeld K (2012) Associations of childhood trauma with hypothalamic-pituitary-adrenal function in borderline personality disorder and major depression. Psychoneuroendocrinology 37(10):1659–1668

Cases O, Seif I, Grimsby J, Gaspar P, Chen K, Pournin S, Müller U, Aguet M, Babinet C, Shih JC, De Maeyer E (1995) Aggressive behavior and altered amounts of brain serotonin and norepinephrine in mice lacking MAOA. Science 268:1763–1766

Casiano H, Belik SL, Cox BJ, Waldman JC, Sareen J (2008) Mental disorder and threats made by noninstitutionalized people with weapons in the national comorbidity survey replication. J Nerv Ment Dis 196:437–445

Cassell MD, Wright DJ (1986) Topography of projections from the medial prefrontal cortex to the amygdala in the rat. Brain Res Bull 17:321–333

Castren E, Elgersma Y, Maffei L, Hagerman R (2012) Treatment of neurodevelopmental disorders in adulthood. J Neurosci 32:14074–14079

Castro WL, Matt KS (1997) The importance of social condition in the hormonal and behavioral responses to an acute social stressor in the male Siberian dwarf hamster (Phodopus sungorus). Horm Behav 32:209–216

Cavigelli SA, Pereira ME (2000) Mating season aggression and fecal testosterone levels in male ring-tailed lemurs (Lemur catta). Horm Behav 37:246–255

Centenaro LA, Vieira K, Zimmermann N, Miczek KA, Lucion AB, de Almeida RM (2008) Social instigation and aggressive behavior in mice: role of 5-HT1A and 5-HT1B receptors in the prefrontal cortex. Psychopharmacology (Berl) 201:237–248

Cercone JJ, Beach SR, Arias I (2005) Gender symmetry in dating intimate partner violence: does similar behavior imply similar constructs? Violence Vict 20:207–218

Challis C, Boulden J, Veerakumar A, Espallergues J, Vassoler FM, Pierce RC, Beck SG, Berton O (2013) Raphe GABAergic neurons mediate the acquisition of avoidance after social defeat. J Neurosci 33:13978–13988

Chamero P, Leinders-Zufall T, Zufall F (2012) From genes to social communication: molecular sensing by the vomeronasal organ. Trends Neurosci 35:597–606

Chang C, Li CY, Earley RL, Hsu YY (2012) Aggression and related behavioral traits: the impact of winning and losing and the role of hormones. Integrative Comp Biol 52:801–813

Chapple CL, Tyler KA, Bersani BE (2005) Child neglect and adolescent violence: examining the effects of self-control and peer rejection. Violence Vict 20:39–53

Cheng SY, Taravosh-Lahn K, Delville Y (2008) Neural circuitry of play fighting in golden hamsters. Neuroscience 156:247–256

Cherek DR (1992) Point subtraction aggression paradigm (PSAP). University of Texas, Houston

Cherek DR, Steinberg JL, Kelly TH, Robinson D (1986) Effects of d-amphetamine on aggressive responding of normal male subjects. Psychiatry Res 21:257–265

Cheu JW, Siegel A (1998) GABA receptor mediated suppression of defensive rage behavior elicited from the medial hypothalamus of the cat: role of the lateral hypothalamus. Brain Res 783:293–304

Chiavegatto S, Quadros IM, Ambar G, Miczek KA (2010) Individual vulnerability to escalated aggressive behavior by a low dose of alcohol: decreased serotonin receptor mRNA in the prefrontal cortex of male mice. Genes Brain Behav 9:110–119

Chichinadze KN, Domianidze TR, Matitaishvili TT, Chichinadze NK, Lazarashvili AG (2010) Possible relation of plasma testosterone level to aggressive behavior of male prisoners. Bull Exp Biol Med 149:7–9

Chitanondh H (1966) Stereotaxic amygdalotomy in the treatment of olfactory seizures and psychiatric disorders with olfactory hallucination. Confin Neurol 27:181–196

Cho JJ, Cadet P, Salamon E, Mantione K, Stefano GB (2003) The nongenomic protective effects of estrogen on the male cardiovascular system: clinical and therapeutic implications in aging men. Med Sci Monit 9:RA63–RA68

Christie MJ, James LB, Beart PM (1986) An excitatory amino acid projection from rat prefrontal cortex to periaqueductal gray. Brain Res Bull 16:127–129

Chu EH, Trosko JE, Chang CC (1977) Mutational approaches to the study of carcinogenesis. J Toxicol Environ Health 2:1317–1334

Chudasama Y, Doobay VM, Liu Y (2012) Hippocampal-prefrontal cortical circuit mediates inhibitory response control in the rat. J Neurosci 32:10915–10924

Cicchetti D, Rogosch FA (2001) The impact of child maltreatment and psychopathology on neuroendocrine functioning. Dev Psychopathol 13:783–804

Cima M, Smeets T, Jelicic M (2008) Self-reported trauma, cortisol levels, and aggression in psychopathic and non-psychopathic prison inmates. Biol Psychol 78:75–86

Clinton SM, Kerman IA, Orr HR, Bedrosian TA, Abraham AD, Simpson DN, Watson SJ, Akil H (2011) Pattern of forebrain activation in high novelty-seeking rats following aggressive encounter. Brain Res 1422:20–31

Clipperton-Allen AE, Almey A, Melichercik A, Allen CP, Choleris E (2011) Effects of an estrogen receptor alpha agonist on agonistic behaviour in intact and gonadectomized male and female mice. Psychoneuroendocrinology 36:981–995

Clutton-Brock TH, Parker GA (1995) Punishment in animal societies. Nature 373:209–216

Coccaro E, McNamee B (1998) Biology of aggression: relevance to crime. In: Skodol AE (ed) Psychopathology and violent crime. American Psychiatric Press, Washington, DC, pp 99–128

Coccaro EF, Bergeman CS, Kavoussi RJ, Seroczynski AD (1997a) Heritability of aggression and irritability: a twin study of the Buss-Durkee aggression scales in adult male subjects. Biol Psychiatry 41:273–284

Coccaro EF, Berman ME, Kavoussi RJ (1997b) Assessment of life history of aggression: development and psychometric characteristics. Psychiatry Res 73:147–157

Coccaro EF, Beresford B, Minar P, Kaskow J, Geracioti T (2007a) CSF testosterone: relationship to aggression, impulsivity, and venturesomeness in adult males with personality disorder. J Psychiatr Res 41:488–492

Coccaro EF, McCloskey MS, Fitzgerald DA, Phan KL (2007b) Amygdala and orbitofrontal reactivity to social threat in individuals with impulsive aggression. Biol Psychiatry 62:168–178

Cohen D, Nisbett RE, Bowdle BF, Schwarz N (1996) Insult, aggression, and the southern culture of honor: an "experimental ethnography". J Pers Soc Psychol 70:945–959

Cohen-Bendahan CC, Buitelaar JK, van Goozen SH, Orlebeke JF, Cohen-Kettenis PT (2005) Is there an effect of prenatal testosterone on aggression and other behavioral traits? A study comparing same-sex and opposite-sex twin girls. Horm Behav 47:230–237

Coker-Appiah DS, White SF, Clanton R, Yang J, Martin A, Blair RJ (2013) Looming animate and inanimate threats: the response of the amygdala and periaqueductal gray. Soc Neurosci 8:621–630

Comoli E, Ribeiro-Barbosa ER, Canteras NS (2003) Predatory hunting and exposure to a live predator induce opposite patterns of Fos immunoreactivity in the PAG. Behav Brain Res 138:17–28

Compaan JC, de Ruiter AJ, Koolhaas JM, van Oortmerssen GA, Bohus B (1992) Differential effects of neonatal testosterone treatment on aggression in two selection lines of mice. Physiol Behav 51:7–10

Compaan JC, Buijs RM, Pool CW, De Ruiter AJ, Koolhaas JM (1993a) Differential lateral septal vasopressin innervation in aggressive and nonaggressive male mice. Brain Res Bull 30:1–6

Compaan JC, van Wattum G, de Ruiter AJ, van Oortmerssen GA, Koolhaas JM, Bohus B (1993b) Genetic differences in female house mice in aggressive response to sex steroid hormone treatment. Physiol Behav 54:899–902

Compaan JC, Hutchison JB, Wozniak A, de Ruiter AJ, Koolhaas JM (1994a) Brain aromatase activity and plasma testosterone levels are elevated in aggressive male mice during early ontogeny. Brain Res Dev Brain Res 82:185–192

Compaan JC, Wozniak A, De Ruiter AJ, Koolhaas JM, Hutchison JB (1994b) Aromatase activity in the preoptic area differs between aggressive and nonaggressive male house mice. Brain Res Bull 35:1–7

Cooper DN (1983) Eukaryotic DNA, methylation. Hum Genet 64:315–333

Cornil CA, Charlier TD (2010) Rapid behavioural effects of oestrogens and fast regulation of their local synthesis by brain aromatase. J Neuroendocrinol 22:664–673

Corum CR, Thurmond JB (1997) Effects of acute exposure to stress on subsequent aggression and locomotion performance. Psychosom Med 39:436–443

Couppis MH, Kennedy CH, Stanwood GD (2008) Differences in aggressive behavior and in the mesocorticolimbic DA system between A/J and BALB/cJ mice. Synapse 62:715–724

Coutinho MR, Menescal-de-Oliveira L (2010) Role of homocysteic acid in the guinea pig (Cavia porcellus) anterior cingulate cortex in tonic immobility and the influence of NMDA receptors on the dorsal PAG. Behav Brain Res 208:237–242

Cox CL, Gotimer K, Roy AK, Castellanos FX, Milham MP, Kelly C (2010) Your resting brain CAREs about your risky behavior. PLoS One 5:e12296

Critchley HD, Simmons A, Daly EM, Russell A, van Amelsvoort T, Robertson DM, Glover A, Murphy DG (2000) Prefrontal and medial temporal correlates of repetitive violence to self and others. Biol Psychiatry 47:928–934

Crowe RR (1974) An adoption study of antisocial personality. Arch Gen Psychiatry 31:785–791

Crowe SL, Blair RJ (2008) The development of antisocial behavior: what can we learn from functional neuroimaging studies? Dev Psychopathol 20:1145–1159

Cullerton-Sen C, Cassidy AR, Murray-Close D, Cicchetti D, Crick NR, Rogosch FA (2008) Childhood maltreatment and the development of relational and physical aggression: the importance of a gender-informed approach. Child Dev 79:1736–1751

Cummings JL (1995) Anatomic and behavioral aspects of frontal-subcortical circuits. Ann NY Acad Sci 769:1–13

Cummings JL, Mega M, Gray K, Rosenberg-Thompson S, Carusi DA, Gornbein J (1994) The neuropsychiatric inventory: comprehensive assessment of psychopathology in dementia. Neurology 44:2308–2314

Cunningham RL, McGinnis MY (2008) Prepubertal social subjugation and anabolic androgenic steroid-induced aggression in male rats. J Neuroendocrinol 20:997–1005

Cunningham RL, Claiborne BJ, McGinnis MY (2007) Pubertal exposure to anabolic androgenic steroids increases spine densities on neurons in the limbic system of male rats. Neuroscience 150:609–615

Cushing BS, Kramer KM (2005) Mechanisms underlying epigenetic effects of early social experience: the role of neuropeptides and steroids. Neurosci Biobehav Rev 29:1089–1105

da Veiga CP, Miczek KA, Lucion AB, de Almeida RM (2011) Social instigation and aggression in postpartum female rats: role of 5-Ht1A and 5-Ht1B receptors in the dorsal raphé nucleus and prefrontal cortex. Psychopharmacology (Berl) 213:475–487

Dabbs JM Jr, Morris R (1990) Testosterone, social class, and antisocial behavior in a sample of 4,462 men. Psychol Sci 1:209–211

Dabbs JM Jr, Jurkovic GJ, Frady RL (1991) Salivary testosterone and cortisol among late adolescent male offenders. J Abnorm Child Psychol 19:469–478

Dalley JW, Theobald DE, Pereira EA, Li PM, Robbins TW (2002) Specific abnormalities in serotonin release in the prefrontal cortex of isolation-reared rats measured during behavioural performance of a task assessing visuospatial attention and impulsivity. Psychopharmacology (Berl) 164:329–340

Damasio AR, Grabowski TJ, Bechara A, Damasio H, Ponto LL, Parvizi J, Hichwa RD (2000) Subcortical and cortical brain activity during the feeling of self-generated emotions. Nat Neurosci 3:1049–1056

Darwin C (1871) The descent of man and selection in relation to sex. Murray, London
Datiche F, Cattarelli M (1996) Reciprocal and topographic connections between the piriform and prefrontal cortices in the rat: a tracing study using the B subunit of the cholera toxin. Brain Res Bull 41:391–398
Datson NA, Morsink MC, Meijer OC, de Kloet ER (2008) Central corticosteroid actions: search for gene targets. Eur J Pharmacol 583:272–289
Davidson RJ, Putnam KM, Larson CL (2000) Dysfunction in the neural circuitry of emotion regulation – a possible prelude to violence. Science 289:591–594
Davis M, Whalen PJ (2001) The amygdala: vigilance and emotion. Mol Psychiatry 6:13–34
Day-Wilson KM, Jones DN, Southam E, Cilia J, Totterdell S (2006) Medial prefrontal cortex volume loss in rats with isolation rearing-induced deficits in prepulse inhibition of acoustic startle. Neuroscience 141:1113–1121
de Almeida RM, Giovenardi M, da Silva SP, de Oliveira VP, Stein DJ (2005) Maternal aggression in Wistar rats: effect of 5-HT2A/2C receptor agonist and antagonist microinjected into the dorsal periaqueductal gray matter and medial septum. Braz J Med Biol Res 38:597–602
de Almeida RM, Rosa MM, Santos DM, Saft DM, Benini Q, Miczek KA (2006) 5-HT(1B) receptors, ventral orbitofrontal cortex, and aggressive behavior in mice. Psychopharmacology (Berl) 185:441–450
de Almeida AN, Fonoff ET, Ballester G, Teixeira MJ, Marino R Jr (2008) Stereotactic disconnection of hypothalamic hamartoma to control seizure and behavior disturbance: case report and literature review. Neurosurg Rev 31:343–349
de Barros DM, de Pádua Serafim A (2008) Association between personality disorder and violent behavior pattern. Forensic Sci Int 179:19–22
de Boer SF, Koolhaas JM (2005) 5-HT1A and 5-HT1B receptor agonists and aggression: a pharmacological challenge of the serotonin deficiency hypothesis. Eur J Pharmacol 526:125–139
de Boer SF, van der Vegt BJ, Koolhaas JM (2003) Individual variation in aggression of feral rodent strains: a standard for the genetics of aggression and violence? Behav Genet 33:485–501
De Brito SA, Mechelli A, Wilke M, Laurens KR, Jones AP, Barker GJ, Hodgins S, Viding E (2009) Size matters: increased grey matter in boys with conduct problems and callous-unemotional traits. Brain 132:843–852
de Bruin JP, van Oyen HG, Van de Poll N (1983) Behavioural changes following lesions of the orbital prefrontal cortex in male rats. Behav Brain Res 10:209–232
de Kloet ER, Han F, Meijer OC (2008a) From the stalk to down under about brain glucocorticoid receptors, stress and development. Neurochem Res 33:637–642
de Kloet ER, Karst H, Joëls M (2008b) Corticosteroid hormones in the central stress response: quick-and-slow. Front Neuroendocrinol 29:268–272
de Oliveira-Souza R, Hare RD, Bramati IE, Garrido GJ, Azevedo Ignácio F, Tovar-Moll F, Moll J (2008) Psychopathy as a disorder of the moral brain: fronto-temporo-limbic grey matter reductions demonstrated by voxel-based morphometry. Neuroimage 40:1202–1213
DeBold JF, Miczek KA (1985) Testosterone modulates the effects of ethanol on male mouse aggression. Psychopharmacology (Berl) 86:286–290
Decety J, Porges EC (2011) Imagining being the agent of actions that carry different moral consequences: an fMRI study. Neuropsychologia 49:2994–3001
Deckel AW, Hesselbrock V, Bauer L (1996) Antisocial personality disorder, childhood delinquency, and frontal brain functioning: EEG and neuropsychological findings. J Clin Psychol 52:639–650
Del Arco A, Mora F (2009) Neurotransmitters and prefrontal cortex-limbic system interactions: implications for plasticity and psychiatric disorders. J Neural Transm 116:941–952
DeLeon KR, Grimes JM, Melloni RH Jr (2002) Repeated anabolic–androgenic steroid treatment during adolescence increases vasopressin V(1A) receptor binding in Syrian hamsters: correlation with offensive aggression. Horm Behav 42:182–191

Delgado-Martínez AD, Vives F (1993) Effects of medial prefrontal cortex stimulation on the spontaneous activity of the ventral pallidal neurons in the rat. Can J Physiol Pharmacol 71:343–347

Dell'Osso B, Berlin HA, Serati M, Altamura AC (2010) Neuropsychobiological aspects, comorbidity patterns and dimensional models in borderline personality disorder. Neuropsychobiology 61:169–179

Delville Y, Mansour KM, Ferris CF (1996) Testosterone facilitates aggression by modulating vasopressin receptors in the hypothalamus. Physiol Behav 60:25–29

Delville Y, Melloni RH Jr, Ferris CF (1998) Behavioral and neurobiological consequences of social subjugation during puberty in golden hamsters. J Neurosci 18:2667–2672

Delville Y, De Vries GJ, Ferris CF (2000) Neural connections of the anterior hypothalamus and agonistic behavior in golden hamsters. Brain Behav Evol 55:53–76

Demas GE, Moffatt CA, Drazen DL, Nelson RJ (1999) Castration does not inhibit aggressive behavior in adult male prairie voles (Microtus ochrogaster). Physiol Behav 66:59–62

Demas GE, Polacek KM, Durazzo A, Jasnow AM (2004) Adrenal hormones mediate melatonin-induced increases in aggression in male Siberian hamsters (Phodopus sungorus). Horm Behav 46:582–591

Demski LS (1973) Feeding and aggressive behavior evoked by hypothalamic stimulation in a cichlid fish. Comp Biochem Physiol A Comp Physiol 44:685–692

DeNardo DF, Sinervo B (1994) Effects of corticosterone on activity and home-range size of free-ranging male lizards. Horm Behav 28:53–65

Denson TF, Pedersen WC, Miller N (2006) The displaced aggression questionnaire. J Pers Soc Psychol 90:1032–10351

Descarries L, Lemay B, Doucet G, Berger B (1987) Regional and laminar density of the dopamine innervation in adult rat cerebral cortex. Neuroscience 21:807–824

DeYoung CG, Hirsh JB, Shane MS, Papademetris X, Rajeevan N, Gray JR (2010) Testing predictions from personality neuroscience. Brain structure and the big five. Psychol Sci 21:820–828

DiBattista JD, Anisman H, Whitehead M, Gilmour KM (2005) The effects of cortisol administration on social status and brain monoaminergic activity in rainbow trout Oncorhynchus mykiss. J Exp Biol 208:2707–2718

Dieckmann G, Schneider-Jonietz B, Schneider H (1988) Psychiatric and neuropsychological findings after stereotactic hypothalamotomy, in cases of extreme sexual aggressivity. Acta Neurochir Suppl 44:163–166

Dinn WM, Harris CL (2000) Neurocognitive function in antisocial personality disorder. Psychiatry Res 97:173–190

Dixson AF (1993) Sexual and aggressive behaviour of adult male marmosets (Callithrix jacchus) castrated neonatally, prepubertally, or in adulthood. Physiol Behav 54:301–307

Dmitrieva TN, Oades RD, Hauffa BP, Eggers C (2001) Dehydroepiandrosterone sulphate and corticotropin levels are high in young male patients with conduct disorder: comparisons for growth factors, thyroid and gonadal hormones. Neuropsychobiology 43:134–140

Dodge KA, Bates JE, Pettit GS (1990) Mechanisms in the cycle of violence. Science 250:1678–1683

Dolan M, Park I (2002) The neuropsychology of antisocial personality disorder. Psychol Med 32:417–427

Dolan M, Anderson IM, Deakin JF (2001) Relationship between 5-HT function and impulsivity and aggression in male offenders with personality disorders. Br J Psychiatry 178:352–359

Dominguez R, Cruz-Morales SE, Carvalho MC, Xavier M, Brandao ML (2003) Effect of steroid injection to newborn rats on serotonin activity in frontal cortex and raphe. Neuroreport 14:597–599

Donegan NH, Sanislow CA, Blumberg HP, Fulbright RK, Lacadie C, Skudlarski P, Gore JC, Olson IR, McGlashan TH, Wexler BE (2003) Amygdala hyperreactivity in borderline personality disorder: implications for emotional dysregulation. Biol Psychiatry 54:1284–1293

Dorn LD, Kolko DJ, Susman EJ, Huang B, Stein H, Music E, Bukstein OG (2009) Salivary gonadal and adrenal hormone differences in boys and girls with and without disruptive behavior disorders: contextual variants. Biol Psychol 81:31–39
Dougherty DD, Shin LM, Alpert NM, Pitman RK, Orr SP, Lasko M, Macklin ML, Fischman AJ, Rauch SL (1999) Anger in healthy men: a PET study using script-driven imagery. Biol Psychiatry 46:466–472
Dowson J, Bazanis E, Rogers R, Prevost A, Taylor P, Meux C, Staley C, Nevison-Andrews D, Taylor C, Robbins T, Sahakian B (2004) Impulsivity in patients with borderline personality disorder. Compr Psychiatry 45:29–36
Dube MG, Kalra SP, Kalra PS (1999) Food intake elicited by central administration of orexins/hypocretins: identification of hypothalamic sites of action. Brain Res 842:473–477
Ducharme S, Hudziak JJ, Botteron KN, Ganjavi H, Lepage C, Collins DL, Albaugh MD, Evans AC, Karama S, Brain Development Cooperative Group (2011) Right anterior cingulate cortical thickness and bilateral striatal volume correlate with child behavior checklist aggressive behavior scores in healthy children. Biol Psychiatry 70:283–290
Duncan GE, Johnson KB, Breese GR (1993) Topographic patterns of brain activity in response to swim stress: assessment by 2-deoxyglucose uptake and expression of Fos-like immunoreactivity. J Neurosci 13:3932–3943
Duncan GE, Inada K, Farrington JS, Koller BH, Moy SS (2009) Neural activation deficits in a mouse genetic model of NMDA receptor hypofunction in tests of social aggression and swim stress. Brain Res 1265:186–195
Earley RL, Edwards JT, Aseem O, Felton K, Blumer LS, Karom M, Grober MS (2006) Social interactions tune aggression and stress responsiveness in a territorial cichlid fish (Archocentrus nigrofasciatus). Physiol Behav 88:353–363
Eckhardt CI, Barbour KA, Davison GC (1998) Articulated irrational thoughts in maritally violent and nonviolent men during anger arousal. J Consulting Clin Psychol 66:259–269
Edinger HM, Siegel A, Troiano R (1975) Effect of stimulation of prefrontal cortex and amygdala on diencephalic neurons. Brain Res 97:17–31
Edwards DA (1969) Early androgen stimulation and aggressive behavior in male and female mice. Physiol Behav 40:333–338
Ehrenkranz J, Bliss E, Sheard MH (1974) Plasma testosterone: correlation with aggressive behavior and social dominance in man. Psychosom Med 36:469–475
Eichelman BS Jr (1971) Effect of subcortical lesions on shock-induced aggression in the rat. J Comp Physiol Psychol 74:331–339
Emery NJ, Capitanio JP, Mason WA, Machado CJ, Mendoza SP, Amaral DG (2001) The effects of bilateral lesions of the amygdala on dyadic social interactions in rhesus monkeys (Macaca mulatta). Behav Neurosci 115:515–544
Ermer E, Cope LM, Nyalakanti PK, Calhoun VD, Kiehl KA (2012) Aberrant paralimbic gray matter in criminal psychopathy. J Abnorm Psychol 121:649–658
Ermer E, Cope LM, Nyalakanti PK, Calhoun VD, Kiehl KA (2013) Aberrant paralimbic gray matter in incarcerated male adolescents with psychopathic traits. J Am Acad Child Adolesc Psychiatry 52:94–103
Essex MJ, Boyce WT, Goldstein LH, Armstrong JM, Kraemer HC, Kupfer DJ, MacArthur Assessment Battery Working Group (2002) The confluence of mental, physical, social, and academic difficulties in middle childhood. II: Developing the Macarthur health and behavior questionnaire. J Am Acad Child Adolesc Psychiatry 41:588–603
Evans SM (1973) A study of fighting reactions in some nereid polychaetes. Anim Behav 21:138–146
Evans DA, Clarke CA (1961) Pharmacogenetics. Br Med Bull 17:234–240
Everts HG, De Ruiter AJ, Koolhaas JM (1997) Differential lateral septal vasopressin in wild-type rats: correlation with aggression. Horm Behav 31:136–144
Fabiansson EC, Denson TF, Moulds ML, Grisham JR, Schira MM (2012) Don't look back in anger: neural correlates of reappraisal, analytical rumination, and angry rumination during recall of an anger-inducing autobiographical memory. Neuroimage 59:2974–2981

Faccidomo S, Quadros IM, Takahashi A, Fish EW, Miczek KA (2012) Infralimbic and dorsal raphé microinjection of the 5-HT(1B) receptor agonist CP-93,129: attenuation of aggressive behavior in CFW male mice. Psychopharmacology (Berl) 222:117–128
Fahim C, He Y, Yoon U, Chen J, Evans A, Pérusse D (2011) Neuroanatomy of childhood disruptive behavior disorders. Aggress Behav 37:326–337
Fairchild G, van Goozen SH, Stollery SJ, Brown J, Gardiner J, Herbert J, Goodyer IM (2008) Cortisol diurnal rhythm and stress reactivity in male adolescents with early-onset or adolescence-onset conduct disorder. Biol Psychiatry 64:599–606
Fairchild G, Hagan CC, Walsh ND, Passamonti L, Calder AJ, Goodyer IM (2013) Brain structure abnormalities in adolescent girls with conduct disorder. J Child Psychol Psychiatry 54:86–95
Fang CY, Egleston BL, Brown KM, Lavigne JV, Stevens VJ, Barton BA, Chandler DW, Dorgan JF (2009) Family cohesion moderates the relation between free testosterone and delinquent behaviors in adolescent boys and girls. J Adolesc Health 44:590–597
Faria MA (2013) Violence, mental illness, and the brain – a brief history of psychosurgery: Part 3 – from deep brain stimulation to amygdalotomy for violent behavior, seizures, and pathological aggression in humans. Surg Neurol Int 4:91
Fava M (1998) Depression with anger attacks. J Clin Psychiatry 59(Suppl 18):18–22
Feek CM, Tuzi NL, Edwards CR (1989) Adrenalectomy does not influence basal secretion of testosterone in rat in vivo. J Steroid Biochem 32:725–728
Ferree ED, Wikelski MC, Anderson DJ (2004) Hormonal correlates of siblicide in Nazca boobies: support for the challenge hypothesis. Horm Behav 46:655–662
Ferris CF (2003) Using an animal model to assess the long-term behavioral and biological consequences of adolescent abuse and exposure to alcohol. Ann NY Acad Sci 1008:69–78
Ferris CF, Delville Y (1994) Vasopressin and serotonin interactions in the control of agonistic behavior. Psychoneuroendocrinology 19:593–601
Ferris CF, Potegal M (1988) Vasopressin receptor blockade in the anterior hypothalamus suppresses aggression in hamsters. Physiol Behav 44:235–239
Ferris CF, Melloni RH, Koppel G, Perry KW, Fuller RW, Delville Y (1997) Vasopressin/serotonin interactions in the anterior hypothalamus control aggressive behavior in golden hamsters. J Neurosci 17:4331–4340
Ferris CF, Messenger T, Sullivan R (2005) Behavioral and neuroendocrine consequences of social subjugation across adolescence and adulthood. Front Zool 2:7
Ferris CF, Stolberg T, Kulkarni P, Murugavel M, Blanchard R, Blanchard DC, Febo M, Brevard M, Simon NG (2008) Imaging the neural circuitry and chemical control of aggressive motivation. BMC Neurosci 9:111
Feshbach S (1971) The function of aggression and the regulation of aggressive drive. Psychol Rev 71:257–272
Fineberg NA, Potenza MN, Chamberlain SR, Berlin HA, Menzies L, Bechara A, Sahakian BJ, Robbins TW, Bullmore ET, Hollander E (2010) Probing compulsive and impulsive behaviors, from animal models to endophenotypes: a narrative review. Neuropsychopharmacology 35:591–604
Finkelstein JW, Susman EJ, Chinchilli VM, Kunselman SJ, D'Arcangelo MR, Schwab J, Demers LM, Liben LS, Lookingbill G, Kulin HE (1997) Estrogen or testosterone increases self-reported aggressive behaviors in hypogonadal adolescents. J Clin Endocrinol Metab 82:2433–2438
Fischer SG, Ricci LA, Melloni RH Jr (2007) Repeated anabolic/androgenic steroid exposure during adolescence alters phosphate-activated glutaminase and glutamate receptor 1 (GluR1) subunit immunoreactivity in Hamster brain: correlation with offensive aggression. Behav Brain Res 180:77–85
Fish EW, Faccidomo S, Miczek KA (1999) Aggression heightened by alcohol or social instigation in mice: reduction by the 5-HT1B receptor agonist CP-94,253. Psychopharmacology (Berl) 146:391–399
Fish EW, DeBold JF, Miczek KA (2005) Escalated aggression as a reward: corticosterone and GABA(A) receptor positive modulators in mice. Psychopharmacology (Berl) 182:116–127

Fitz D (1976) A renewed look at Miller's conflict theory of aggression displacement. J Pers Soc Psychol 33:725–732
Floyd NS, Price JL, Ferry AT, Keay KA, Bandler R (2001) Orbitomedial prefrontal cortical projections to hypothalamus in the rat. J Comp Neurol 432:307–328
Fokidis HB, Prior NH, Soma KK (2013) Fasting increases aggression and differentially modulates local and systemic steroid levels in male zebra finches. Endocrinology 154:4328–4339
Follenius M, Simon C, Brandenberger G, Lenzi P (1987) Ultradian plasma corticotropin and cortisol rhythms: time-series analyses. J Endocrinol Invest 10:261–266
Foradori CD, Weiser MJ, Handa RJ (2008) Non-genomic actions of androgens. Front Neuroendocrinol 29:169–181
Fountas KN, Smith JR (2007) Historical evolution of stereotactic amygdalotomy for the management of severe aggression. J Neurosurg 106:710–713
Fowler KA, Westen D (2011) Subtyping male perpetrators of intimate partner violence. J Interpers Violence 26:607–639
Frahm KA, Lumia AR, Fernandez E, Strong R, Roberts JL, McGinnis MY (2011) Effects of anabolic androgenic steroids and social subjugation on behavior and neurochemistry in male rats. Pharmacol Biochem Behav 97:416–422
Frankle WG, Lombardo I, New AS, Goodman M, Talbot PS, Huang Y, Hwang DR, Slifstein M, Curry S, Abi-Dargham A, Laruelle M, Siever LJ (2005) Brain serotonin transporter distribution in subjects with impulsive aggressivity: a positron emission study with [11C]McN 5652. Am J Psychiatry 162:915–923
Franklin KBJ, Paxinos G (2001) The mouse brain in stereotaxic coordinates. Academic, San Diego, CA
Franzini A, Marras C, Ferroli P, Bugiani O, Broggi G (2005) Stimulation of the posterior hypothalamus for medically intractable impulsive and violent behavior. Stereotact Funct Neurosurg 83:63–66
Franzini A, Marras C, Ferroli P, Bugiani O, Broggi G (2008) Stimulation of the posterior hypothalamus for medically intractable impulsive and violent behavior. Stereotact Funct Neurosurg 83:63–66
Franzini A, Messina G, Cordella R, Marras C, Broggi G (2010) Deep brain stimulation of the posteromedial hypothalamus: indications, long-term results, and neurophysiological considerations. Neurosurg Focus 29:E13
Fuchs SA, Edinger HM, Siegel A (1985) The organization of the hypothalamic pathways mediating affective defense behavior in the cat. Brain Res 330:77–92
Fukui H, Murai T, Shinozaki J, Aso T, Fukuyama H, Hayashi T, Hanakawa T (2006) The neural basis of social tactics: an fMRI study. Neuroimage 32:913–920
Fulwiler CE, King JA, Zhang N (2012) Amygdala-orbitofrontal resting-state functional connectivity is associated with trait anger. Neuroreport 23:606–610
Fusar-Poli P, Placentino A, Carletti F, Landi P, Allen P, Surguladze S, Benedetti F, Abbamonte M, Gasparotti R, Barale F, Perez J, McGuire P, Politi P (2009) Functional atlas of emotional faces processing: a voxel-based meta-analysis of 105 functional magnetic resonance imaging studies. J Psychiatry Neurosci 34:418–432
Fuxjager MJ, Montgomery JL, Marler CA (2011) Species differences in the winner effect disappear in response to post-victory testosterone manipulations. Proc Biol Sci 278:3497–3503
Gaab J, Rohleder N, Nater UM, Ehlert U (2005) Psychological determinants of the cortisol stress response: the role of anticipatory cognitive appraisal. Psychoneuroendocrinology 30:599–610
Gabbott PL, Warner TA, Jays PR, Bacon SJ (2003) Areal and synaptic interconnectivity of prelimbic (area 32), infralimbic (area 25) and insular cortices in the rat. Brain Res 993:59–71
Gabbott PL, Warner TA, Jays PR, Salway P, Busby SJ (2005) Prefrontal cortex in the rat: projections to subcortical autonomic, motor and limbic centres. J Comp Neurol 492:145–177
Gabor CS, Phan A, Clipperton-Allen AE, Kavaliers M, Choleris E (2012) Interplay of oxytocin, vasopressin, and sex hormones in the regulation of social recognition. Behav Neurosci 126:97–109

Galen B, Underwood M (1997) A developmental investigation of social aggression among children. Development Psychol 33:589–600

Gammie SC, Nelson RJ (2001) cFOS and pCREB activation and maternal aggression in mice. Brain Res 898:232–241

Gammie SC, Stevenson SA (2006) Effects of daily and acute restraint stress during lactation on maternal aggression and behavior in mice. Stress 9:171–180

Gammie SC, Hasen NS, Stevenson SA, Bale TL, D'Anna KL (2005) Elevated stress sensitivity in corticotropin-releasing factor receptor 2 deficient mice decreases maternal, but not intermale aggression. Behav Brain Res 160:169–177

Gandelman R (1972) Mice: postpartum aggression elicited by the presence of an intruder. Horm Behav 3:23–28

Gandelman R, Simon NG, McDermott NJ (1979) Prenatal exposure to testosterone and its precursors influences morphology and later behavioral responsiveness to testosterone of female mice. Physiol Behav 23:23–26

Gansler DA, Lee AK, Emerton BC, D'Amato C, Bhadelia R, Jerram M, Fulwiler C (2011) Prefrontal regional correlates of self-control in male psychiatric patients: impulsivity facets and aggression. Psychiatry Res 191:16–23

Gao Y, Raine A (2010) Successful and unsuccessful psychopaths: a neurobiological model. Behav Sci Law 28:194–210

Gao HB, Shan LX, Monder C, Hardy MP (1996) Suppression of endogenous corticosterone levels in vivo increases the steroidogenic capacity of purified rat Leydig cells in vitro. Endocrinology 137:1714–1718

Gao Y, Glenn AL, Schug RA, Yang Y, Raine A (2009) The neurobiology of psychopathy: a neurodevelopmental perspective. Can J Psychiatry 54:813–823

Geniole SN, Carré JM, McCormick CM (2011) State, not trait, neuroendocrine function predicts costly reactive aggression in men after social exclusion and inclusion. Biol Psychol 87:137–145

George DT, Umhau JC, Phillips MJ, Emmela D, Ragan PW, Shoaf SE, Rawlings RR (2001) Serotonin, testosterone and alcohol in the etiology of domestic violence. Psychiatry Res 104:27–37

George DT, Rawlings RR, Williams WA, Phillips MJ, Fong G, Kerich M, Momenan R, Umhau JC, Hommer D (2004) A select group of perpetrators of domestic violence: evidence of decreased metabolism in the right hypothalamus and reduced relationships between cortical/subcortical brain structures in position emission tomography. Psychiatry Res 130:11–25

Gerra G, Zaimovic A, Avanzini P, Chittolini B, Giucastro G, Caccavari R, Palladino M, Maestri D, Monica C, Delsignore R, Brambilla F (1997) Neurotransmitter-neuroendocrine responses to experimentally induced aggression in humans: influence of personality variable. Psychiatry Res 66:33–43

Gerra G, Zaimovic A, Ampollini R, Giusti F, Delsignore R, Raggi MA, Laviola G, Macchia T, Brambilla F (2001a) Experimentally induced aggressive behavior in subjects with 3,4-methylenedioxy-methamphetamine ("Ecstasy") use history: psychobiological correlates. J Subst Abuse 13:471–491

Gerra G, Zaimovic A, Raggi MA, Giusti F, Delsignore R, Bertacca S, Brambilla F (2001b) Aggressive responding of male heroin addicts under methadone treatment: psychometric and neuroendocrine correlates. Drug Alcohol Depend 65:85–95

Gerra G, Zaimovic A, Moi G, Bussandri M, Bubici C, Mossini M, Raggi MA, Brambilla F (2004) Aggressive responding in abstinent heroin addicts: neuroendocrine and personality correlates. Prog Neuropsychopharmacol Biol Psychiatry 28:129–139

Gerra G, Zaimovic A, Raggi MA, Moi G, Branchi B, Moroni M, Brambilla F (2007) Experimentally induced aggressiveness in heroin-dependent patients treated with buprenorphine: comparison of patients receiving methadone and healthy subjects. Psychiatry Res 149:201–213

Gilpin NW, Roberto M (2012) Neuropeptide modulation of central amygdala neuroplasticity is a key mediator of alcohol dependence. Neurosci Biobehav Rev 36:873–888

Gleason ED, Fuxjager MJ, Oyegbile TO, Marler CA (2009) Testosterone release and social context: when it occurs and why. Front Neuroendocrinol 30:460–469

Glenn AL, Raine A, Schug RA (2009) The neural correlates of moral decision-making in psychopathy. Mol Psychiatry 14:5–6

Glenn AL, Raine A, Schug RA, Gao Y, Granger DA (2011) Increased testosterone-to-cortisol ratio in psychopathy. J Abnorm Psychol 120:389–399

Gnyszka A, Jastrzebski Z, Flis S (2013) DNA methyltransferase inhibitors and their emerging role in epigenetic therapy of cancer. Anticancer Res 33:2989–2996

Gobrogge KL, Liu Y, Jia X, Wang Z (2007) Anterior hypothalamic neural activation and neurochemical associations with aggression in pair-bonded male prairie voles. J Comp Neurol 502:1109–1122

Gobrogge KL, Liu Y, Young LJ, Wang Z (2009) Anterior hypothalamic vasopressin regulates pair-bonding and drug-induced aggression in a monogamous rodent. Proc Natl Acad Sci U S A 106:19144–19149

Goethals I, Audenaert K, Jacobs F, Van den Eynde F, Bernagie K, Kolindou A, Vervaet M, Dierckx R, Van Heeringen C (2005) Brain perfusion SPECT in impulsivity-related personality disorders. Behav Brain Res 157:187–192

Gold M, Gefter M, Hausmann R, Hurwitz J (1966) Methylation of DNA. J Gen Physiol 49:5–28

Golden CJ, Jackson ML, Peterson-Rohne A, Gontkovsky ST (1996) Neuropsychological correlates of violence and aggression: a review of the clinical literature. Aggress Violent Behav 1(1):3–25

Goldman-Rakic PS (1996) The prefrontal landscape: implications of functional architecture for understanding human mentation and the central executive. Philos Trans R Soc Lond B Biol Sci 351:1445–1453

Goltz F (1892) Der Hund ohne Grosshirn. Siebente Abhandlung über die Verrichtungen des Grosshirns. Arch f d ges Physiol 51:570–614, Cited based on Fountas and Smith, 2007

Golubchik P, Mozes T, Maayan R, Weizman A (2009) Neurosteroid blood levels in delinquent adolescent boys with conduct disorder. Eur Neuropsychopharmacol 19:49–52

Gomes MM (2007) A concept analysis of relational aggression. J Psychiatr Ment Health Nurs 14:510–515

Goodall J (1986) Social rejection, exclusion, and shunning among the Gombe chimpanzees. Ethol Sociobiol 7:227–236

Gopal A, Clark E, Allgair A, D'Amato C, Furman M, Gansler DA, Fulwiler C (2013) Dorsal/ventral parcellation of the amygdala: relevance to impulsivity and aggression. Psychiatry Res 211:24–30

Gordis EB, Granger DA, Susman EJ, Trickett PK (2006) Asymmetry between salivary cortisol and alpha-amylase reactivity to stress: relation to aggressive behavior in adolescents. Psychoneuroendocrinology 31:976–987

Gould TJ (2010) Addiction and cognition. Addict Sci Clin Pract 5:4–14

Goyer PF, Andreason PJ, Semple WE, Clayton AH, King AC, Compton-Toth BA, Schulz SC, Cohen RM (1994) Positron-emission tomography and personality disorders. Neuropsychopharmacology 10:21–28

Goymann W, Landys MM, Wingfield JC (2007) Distinguishing seasonal androgen responses from male-male androgen responsiveness-revisiting the challenge hypothesis. Horm Behav 51:463–476

Grant EC, Mackintosh JH (1963) A comparison of the social postures of some common laboratory rodents. Behaviour 21:246–259

Gratton A, Wise RA (1988) Comparisons of refractory periods for medial forebrain bundle fibers subserving stimulation-induced feeding and brain stimulation reward: a psychophysical study. Brain Res 438:256–263

Greenberg N, Scott M, Crews D (1984) Role of the amygdala in the reproductive and aggressive behavior of the lizard, Anolis carolinensis. Physiol Behav 32:147–151

Gregg TR, Siegel A (2001) Brain structures and neurotransmitters regulating aggression in cats: implications for human aggression. Prog Neuropsychopharmacol Biol Psychiatry 25:91–140

Gregg TR, Siegel A (2003) Differential effects of NK1 receptors in the midbrain periaqueductal gray upon defensive rage and predatory attack in the cat. Brain Res 994:55–66
Gregory S, Ffytche D, Simmons A, Kumari V, Howard M, Hodgins S, Blackwood N (2012) The antisocial brain: psychopathy matters. Arch Gen Psychiatry 69:962–972
Grimes JM, Melloni RH Jr (2002) Serotonin modulates offensive attack in adolescent anabolic steroid-treated hamsters. Pharmacol Biochem Behav 73:713–721
Grimes JM, Melloni RH Jr (2005) Serotonin-1B receptor activity and expression modulate the aggression-stimulating effects of adolescent anabolic steroid exposure in hamsters. Behav Neurosci 119:1184–1194
Grimes JM, Melloni RH Jr (2006) Prolonged alterations in the serotonin neural system following the cessation of adolescent anabolic-androgenic steroid exposure in hamsters (Mesocricetus auratus). Behav Neurosci 120:1242–1251
Grimes JM, Ricci LA, Melloni RH Jr (2003) Glutamic acid decarboxylase (GAD65) immunoreactivity in brains of aggressive, adolescent anabolic steroid-treated hamsters. Horm Behav 44:271–280
Groeneweg FL, Karst H, de Kloet ER, Joëls M (2011) Rapid non-genomic effects of corticosteroids and their role in the central stress response. J Endocrinol 209:153–167
Groenewegen HJ (1988) Organization of the afferent connections of the mediodorsal thalamic nucleus in the rat, related to the mediodorsal-prefrontal topography. Neuroscience 24:379–431
Guillot PV, Sluyter F, Crusio WE, Chapouthier G (1999) Mice selected for differences in sensitivity to a benzodiazepine receptor inverse agonist vary in intermale aggression. Neurogenetics 2:171–175
Gunnar MR, Vazquez DM (2001) Low cortisol and a flattening of expected daytime rhythm: potential indices of risk in human development. Develop Psychopathol 13:515–538
Gunnar MR, Kryzer E, Van Ryzin MJ, Phillips DA (2010) The rise in cortisol in family day care: associations with aspects of care quality, child behavior, and child sex. Child Dev 8:851–869
Guth W, Schmittberger R, Schwarze B (1982) An experimental analysis of ultimatum bargaining. J Econ Behav Organ 3:367–388
Guzman-Ramos K, Bermudez-Rattoni F (2012) Interplay of amygdala and insular cortex during and after associative taste aversion memory formation. Rev Neurosci 23:463–471
Haffmans PM, Sival RC, Lucius SA, Cats Q, van Gelder L (2001) Bright light therapy and melatonin in motor restless behaviour in dementia: a placebo-controlled study. Int J Geriatr Psychiatry 16:106–110
Hahn-Holbrook J, Holt-Lunstad J, Holbrook C, Coyne SM, Lawson ET (2011) Maternal defense: breast feeding increases aggression by reducing stress. Psychol Sci 22:1288–1295
Halasz J, Liposits Z, Kruk MR, Haller J (2002a) Neural background of glucocorticoid dysfunction-induced abnormal aggression in rats: involvement of fear- and stress-related structures. Eur J Neurosci 15:561–569
Halasz J, Liposits Z, Meelis W, Kruk MR, Haller J (2002b) Hypothalamic attack area-mediated activation of the forebrain in aggression. Neuroreport 13:1267–1270
Halász J, Tóth M, Kalló I, Liposits Z, Haller J (2006) The activation of prefrontal cortical neurons in aggression – a double labeling study. Behav Brain Res 175:166–175
Halasz J, Toth M, Mikics E, Hrabovszky E, Barsy B, Barsvari B, Haller J (2008) The effect of neurokinin1 receptor blockade on territorial aggression and in a model of violent aggression. Biol Psychiatry 63:271–278
Halasz J, Zelena D, Toth M, Tulogdi A, Mikics E, Haller J (2009) Substance P neurotransmission and violent aggression: the role of tachykinin NK(1) receptors in the hypothalamic attack area. Eur J Pharmacol 611:35–43
Halgren E (1982) Mental phenomena induced by stimulation in the limbic system. Hum Neurobiol 1:251–260
Hallberg M, Johansson P, Kindlundh AM, Nyberg F (2000) Anabolic-androgenic steroids affect the content of substance P and substance P(1–7) in the rat brain. Peptides 21:845–852
Haller J (1995) Biochemical background for an analysis of cost-benefit interrelations in aggression. Neurosci Biobehav Rev 19:599–604

Haller J (2013) The neurobiology of abnormal manifestations of aggression–a review of hypothalamic mechanisms in cats, rodents, and humans. Brain Res Bull 93:97–109
Haller J, Kruk MR (2006) Normal and abnormal aggression: human disorders and novel laboratory models. Neurosci Biobehav Rev 30:292–303
Haller J, Barna I, Baranyi M (1995) Hormonal and metabolic responses during psychosocial stimulation in aggressive and nonaggressive rats. Psychoneuroendocrinology 20:65–74
Haller J, Kiem DT, Makara GB (1996) The physiology of social conflict in rats: what is particularly stressful? Behav Neurosci 110:353–359
Haller J, Albert I, Makara GB (1997) Acute behavioural effects of corticosterone lack specificity but show marked context-dependency. J Neuroendocrinol 9:515–518
Haller J, Abrahám I, Zelena D, Juhász G, Makara GB, Kruk MR (1998a) Aggressive experience affects the sensitivity of neurons towards pharmacological treatment in the hypothalamic attack area. Behav Pharmacol 9:469–475
Haller J, Halasz J, Makara GB, Kruk MR (1998b) Acute effects of glucocorticoids: behavioral and pharmacological perspectives. Neurosci Biobehav Rev 23:337–344
Haller J, Millar S, Kruk MR (1998c) Mineralocorticoid receptor blockade inhibits aggressive behaviour in male rats. Stress 2:201–207
Haller J, Halasz J, Mikics E, Kruk MR, Makara GB (2000a) Ultradian corticosterone rhythm and the propensity to behave aggressively in male rats. J Neuroendocrinol 12:937–940
Haller J, Millar S, van de Schraaf J, de Kloet RE, Kruk MR (2000b) The active phase-related increase in corticosterone and aggression are linked. J Neuroendocrinol 12:431–436
Haller J, van de Schraaf J, Kruk MR (2001) Deviant forms of aggression in glucocorticoid hyporeactive rats: a model for "pathological" aggression? J Neuroendocrinol 13:102–107
Haller J, Halász J, Mikics E, Kruk MR (2004) Chronic glucocorticoid deficiency-induced abnormal aggression, autonomic hypoarousal, and social deficit in rats. J Neuroendocrinol 16:550–557
Haller J, Mikics E, Halász J, Tóth M (2005a) Mechanisms differentiating normal from abnormal aggression: glucocorticoids and serotonin. Eur J Pharmacol 526:89–100
Haller J, Toth M, Halasz J (2005b) The activation of raphe serotonergic neurons in normal and hypoarousal-driven aggression: a double labeling study in rats. Behav Brain Res 161:88–94
Haller J, Toth M, Halasz J, De Boer SF (2006) Patterns of violent aggression-induced brain c-fos expression in male mice selected for aggressiveness. Physiol Behav 88:173–182
Haller J, Horváth Z, Bakos N (2007) The effect of buspirone on normal and hypoarousal-driven abnormal aggression in rats. Prog Neuropsychopharmacol Biol Psychiatry 31:27–31
Haller J, Mikics E, Makara GB (2008) The effects of non-genomic glucocorticoid mechanisms on bodily functions and the central neural system. A critical evaluation of findings. Front Neuroendocrinol 29:273–291
Hammack SE, Cooper MA, Lezak KR (2012) Overlapping neurobiology of learned helplessness and conditioned defeat: implications for PTSD and mood disorders. Neuropharmacology 62:565–575
Hampson E, Ellis CL, Tenk CM (2008) On the relation between 2D:4D and sex-dimorphic personality traits. Arch Sex Behav 37:133–144
Han X, Boyden ES (2007) Multiple-color optical activation, silencing, and desynchronization of neural activity, with single-spike temporal resolution. PLoS One 2:e299
Han Y, Shaikh MB, Siegel A (1996) Medial amygdaloid suppression of predatory attack behavior in the cat: I. Role of a substance P pathway from the medial amygdala to the medial hypothalamus. Brain Res 716:59–71
Hanakawa T (2011) Rostral premotor cortex as a gateway between motor and cognitive networks. Neurosci Res 70:144–154
Harrison RJ, Connor DF, Nowak C, Nash K, Melloni RH Jr (2000) Chronic anabolic–androgenic steroid treatment during adolescence increases anterior hypothalamic vasopressin and aggression in intact hamsters. Psychoneuroendocrinology 25:317–338
Harz H, Hegemann P (1991) Rhodopsin-regulated calcium currents in Chlamydomonas. Nature 351:489–491

Hasen NS, Gammie SC (2005) Differential fos activation in virgin and lactating mice in response to an intruder. Physiol Behav 84:681–695
Hasen NS, Gammie SC (2006) Maternal aggression: new insights from Egr-1. Brain Res 1108:147–156
Hasen NS, Gammie SC (2009) Trpc2 gene impacts on maternal aggression, accessory olfactory bulb anatomy and brain activity. Genes Brain Behav 8:639–649
Hassanain M, Bhatt S, Siegel A (2003) Differential modulation of feline defensive rage behavior in the medial hypothalamus by 5-HT1A and 5-HT2 receptors. Brain Res 981:201–209
Hassanain M, Bhatt S, Zalcman S, Siegel A (2005) Potentiating role of interleukin-1beta (IL-1beta) and IL-1beta type 1 receptors in the medial hypothalamus in defensive rage behavior in the cat. Brain Res 1048:1–11
Haug M, Spetz JF, Schlegel ML, Robel P (1983) Dehydrogriandrosterone inhibits aggressive behavior in castrated male mice. C R Seances Acad Sci III 296:975–977
Haug M, Simler S, Ciesielski L, Mandel P, Moutier R (1984) Influence of castration and brain GABA levels in three strains of mice on aggression towards lactating intruders. Physiol Behav 32:767–770
Hayden-Hixson DM, Ferris CF (1991a) Cortisol exerts site-, context- and dose-dependent effects on agonistic responding in hamsters. J Neuroendocrinol 3:613–622
Hayden-Hixson DM, Ferris CF (1991b) Steroid specific regulation of agonistic responding in the anterior hypothalamus of male hamsters. Physiol Behav 50:793–799
Hazlett EA, New AS, Newmark R, Haznedar MM, Lo JN, Speiser LJ, Chen AD, Mitropoulou V, Minzenberg M, Siever LJ, Buchsbaum MS (2005) Reduced anterior and posterior cingulate gray matter in borderline personality disorder. Biol Psychiatry 58:614–623
He F, Tai F, Zhang Y, Zhang X (2012) Effects of castration on aggression and levels of serum sex hormones and their central receptors in mandarin voles (Microtus mandarinus). J Comp Physiol A Neuroethol Sens Neural Behav Physiol 198:347–362
Heekeren HR, Wartenburger I, Schmidt H, Schwintowski HP, Villringer A (2003) An fMRI study of simple ethical decision-making. Neuroreport 14:1215–1219
Heide KM, Solomon EP (2009) Female juvenile murderers: biological and psychological dynamics leading to homicide. Int J Law Psychiatry 32:244–252
Heimburger RF, Whitlock CC, Kalsbeck JE (1966) Stereotaxic amygdalotomy for epilepsy with aggressive behavior. JAMA 198:741–745
Heimer L, de Olmos JS, Alheid GF, Pearson J, Sakamoto N, Shinoda K, Marksteiner J, Switzer RC III (1999) The human basal forebrain, Part II. In: Bloom FE, Björklund A, Hökfelt T (eds) The primate nervous system, Part III, vol 15, Handbook of chemical neuroanatomy. Elsevier, Amsterdam, pp 57–226
Heinlein CA, Chang C (2002) The roles of androgen receptors and androgen-binding proteins in nongenomic androgen actions. Mol Endocrinol 16:2181–2187
Herberg LJ, Blundell JE (1967) Lateral hypothalamus: hoarding behavior elicited by electrical stimulation. Science 155:349–350
Herman BH (1990) A possible role of proopiomelanocortin peptides in self-injurious behavior. Prog Neuropsychopharmacol Biol Psychiatry 14(Suppl):S109–S139
Herman JP, Flak J, Jankord R (2008) Chronic stress plasticity in the hypothalamic paraventricular nucleus. Prog Brain Res 170:353–364
Hermans EJ, Ramsey NF, van Honk J (2008) Exogenous testosterone enhances responsiveness to social threat in the neural circuitry of social aggression in humans. Biol Psychiatry 63:263–270
Hernando V, Pastor J, Pedrosa M, Peña E, Sola RG (2008) Low-frequency bilateral hypothalamic stimulation for treatment of drug-resistant aggressiveness in a young man with mental retardation. Stereotact Funct Neurosurg 86:219–223
Herpertz SC, Sass H (2000) Emotional deficiency and psychopathy. Behav Sci Law 18:567–580
Herpertz SC, Dietrich TM, Wenning B, Krings T, Erberich SG, Willmes K, Thron A, Sass H (2001) Evidence of abnormal amygdala functioning in borderline personality disorder: a functional MRI study. Biol Psychiatry 50:292–298

Herpertz SC, Huebner T, Marx I, Vloet TD, Fink GR, Stoecker T, Shah NJ, Konrad K, Herpertz-Dahlmann B (2008) Emotional processing in male adolescents with childhood-onset conduct disorder. J Child Psychol Psychiatry 49:781–791

Hervé D, Pickel VM, Joh TH, Beaudet A (1987) Serotonin axon terminals in the ventral tegmental area of the rat: fine structure and synaptic input to dopaminergic neurons. Brain Res 435:71–83

Hess WR (1928) Stammganglien-Reizversuche. Berichte Gesamt Physiol 42:554–555

Hill R, Simpson B, Millet G, Manning J, Kilduff L (2012) Right-left digit ratio (2D:4D) and maximal oxygen uptake. J Sports Sci 30:129–134

Hinde RA (1974) Biological bases of human social behaviour. McGraw-Hill, New York, NY

Hines M (2006) Prenatal testosterone and gender-related behaviour. Eur J Endocrinol 155(Suppl 1):S115–S121

Hines M (2010) Sex-related variation in human behavior and the brain. Trends Cogn Sci 14:448–456

Hines M, Golombok S, Rust J, Johnston KJ, Golding J, Avon Longitudinal Study of Parents and Children Study Team (2002) Testosterone during pregnancy and gender role behavior of preschool children: a longitudinal, population study. Child Dev 73:1678–1687

Hirono N, Mega MS, Dinov ID, Mishkin F, Cummings JL (2000) Left frontotemporal hypoperfusion is associated with aggression in patients with dementia. Arch Neurol 57:861–866

Hitchcock E, Cairns V (1973) Amygdalotomy. Postgrad Med J 49:894–904

Hoaken PN, Giancola PR, Pihl RO (1998) Executive cognitive functions as mediators of alcohol-related aggression. Alcohol Alcohol 33:47–54

Hoebel BG, Teitelbaum P (1962) Hypothalamic control of feeding and self-stimulation. Science 135:375–37

Hoffmann P, Thorén P, Ely D (1987) Effect of voluntary exercise on open-field behavior and on aggression in the spontaneously hypertensive rat (SHR). Behav Neural Biol 47:346–355

Hofmann HD, Seidl K, Unsicker K (1989) Development and plasticity of adrenal chromaffin cells: cues based on in vitro studies. J Electron Microsc Tech 12:397–407

Höglund E, Balm PH, Winberg S (2002) Behavioural and neuroendocrine effects of environmental background colour and social interaction in Arctic charr (Salvelinus alpinus). J Exp Biol 205:2535–2543

Holi M, Auvinen-Lintunen L, Lindberg N, Tani P, Virkkunen M (2006) Inverse correlation between severity of psychopathic traits and serum cortisol levels in young adult violent male offenders. Psychopathology 39:102–104

Holm O (1982) The effect of intent, reason and harm on attribution of aggressiveness. J Psychol 110:49–52

Homberg JR, Pattij T, Janssen MC, Ronken E, De Boer SF, Schoffelmeer AN, Cuppen E (2007) Serotonin transporter deficiency in rats improves inhibitory control but not behavioural flexibility. Eur J Neurosci 26:2066–2073

Hood TW, Siegfried J, Wieser HG (1983) The role of stereotactic amygdalotomy in the treatment of temporal lobe epilepsy associated with behavioral disorders. Appl Neurophysiol 46:19–25

Hoover WB, Vertes RP (2011) Projections of the medial orbital and ventral orbital cortex in the rat. J Comp Neurol 519:3766–3801

Hoshi E, Tanji J (2007) Distinctions between dorsal and ventral premotor areas: anatomical connectivity and functional properties. Curr Opin Neurobiol 17:234–242

Hrabovszky E, Halász J, Meelis W, Kruk MR, Liposits Z, Haller J (2005) Neurochemical characterization of hypothalamic neurons involved in attack behavior: glutamatergic dominance and co-expression of thyrotropin-releasing hormone in a subset of glutamatergic neurons. Neuroscience 133:657–666

Hutchison JB (1974) Post-castration decline in behavioural responsiveness in intrahypothalamic androgen in doves. Brain Res 81:169–181

Hyde JS (2014) Gender similarities and differences. Annu Rev Psychol 65:373–398

Ing NH (2005) Steroid hormones regulate gene expression posttranscriptionally by altering the stabilities of messenger RNAs. Biol Reprod 72:1290–1296

Ivanov I, Yehuda R, Greenblatt E, Davidow J, Makotkine I, Alfi L, Newcorn JH (2011) The effect of trauma on stress reactivity in aggressive youth. Psychiatry Res 189:396–402

Izquierdo A (2005) Comparison of the effects of bilateral orbital prefrontal cortex lesions and amygdala lesions on emotional responses in rhesus monkeys. J Neurosci 25:8534–8542

Jackson D, Burns R, Trksak G, Simeone B, DeLeon KR, Connor DF, Harrison RJ, Melloni RH Jr (2005) Anterior hypothalamic vasopressin modulates the aggression-stimulating effects of adolescent cocaine exposure in Syrian hamsters. Neuroscience 133:635–646

Jacomy H, Burlet A, Bosler O (1999) Vasoactive intestinal peptide neurons as synaptic targets for vasopressin neurons in the suprachiasmatic nucleus. Double-label immunocytochemical demonstration in the rat. Neuroscience 88:859–870

Jasnow AM, Huhman KL, Bartness TJ, Demas GE (2000) Short-day increases in aggression are inversely related to circulating testosterone concentrations in male Siberian hamsters (Phodopus sungorus). Horm Behav 38:102–110

Jasnow AM, Huhman KL, Bartness TJ, Demas GE (2002) Short days and exogenous melatonin increase aggression of male Syrian hamsters (Mesocricetus auratus). Horm Behav 42:13–20

Jasper HH, Marsan CA (1960) A stereotaxic atlas of the diencephalon of the cat. National Research Council of Canada, Ottawa

Joels M, de Kloet ER (1992) Control of neuronal excitability by corticosteroid hormones. Trends Neurosci 15:25–30

Joëls M, Karst H, Alfarez D, Heine VM, Qin Y, van Riel E, Verkuyl M, Lucassen PJ, Krugers HJ (2004) Effects of chronic stress on structure and cell function in rat hippocampus and hypothalamus. Stress 7:221–231

Johansson P, Hallberg M, Kindlundh A, Nyberg F (2000) The effect on opioid peptides in the rat brain, after chronic treatment with the anabolic androgenic steroid, nandrolone decanoate. Brain Res Bull 51:413–418

Johnsen TS, Zuk M (1995) Testosterone and aggression in male red jungle fowl. Horm Behav 29:593–598

Jones AP, Laurens KR, Herba CM, Barker GJ, Viding E (2009) Amygdala hypoactivity to fearful faces in boys with conduct problems and callous-unemotional traits. Am J Psychiatry 166:95–102

Joppa MA, Meisel RL, Garber MA (1995) Fos expression in female hamster brain following sexual and aggressive behaviors. Neuroscience 68:783–792

Juengling FD, Schmahl C, Hesslinger B, Ebert D, Bremner JD, Gostomzyk J, Bohus M, Lieb K (2003) Positron emission tomography in female patients with borderline personality disorder. J Psychiatr Res 37:109–115

Juhász C, Behen ME, Muzik O, Chugani DC, Chugani HT (2001) Bilateral medial prefrontal and temporal neocortical hypometabolism in children with epilepsy and aggression. Epilepsia 42:991–1001

Kaiser S, Harderthauer S, Sachser N, Hennessy MB (2007) Social housing conditions around puberty determine later changes in plasma cortisol levels and behavior. Physiol Behav 90:405–411

Kalnin AJ, Edwards CR, Wang Y, Kronenberger WG, Hummer TA, Mosier KM, Dunn DW, Mathews VP (2011) The interacting role of media violence exposure and aggressive-disruptive behavior in adolescent brain activation during an emotional Stroop task. Psychiatry Res 192:12–19

Kalra PS, Kalra SP (1977) Circadian periodicities of serum androgens, progesterone, gonadotropins and luteinizing hormone-releasing hormone in male rats: the effects of hypothalamic deafferentation, castration and adrenalectomy. Endocrinology 101:1821–1827

Kamigaki T, Fukushima T, Tamura K, Miyashita Y (2012) Neurodynamics of cognitive set shifting in monkey frontal cortex and its causal impact on behavioral flexibility. J Cogn Neurosci 24:2171–2185

Kariyawasam SH, Zaw F, Handley SL (2002) Reduced salivary cortisol in children with comorbid attention deficit hyperactivity disorder and oppositional defiant disorder. Neuro Endocrinol Lett 23:45–48

Karli PM, Vergnes M, Eclancher F, Schmitt P, Chawerand IP (1972) Role of amygdala in the control of mouse-killing behavior in the rat. Adv Behav Biol 2:533–580

Kaufmann H, Feshbach S (1963) Displaced aggression and its modification through exposure to antiaggressive communications. J Abnorm Soc Psychol 67:79–83

Kavushansky A, Ben-Shachar D, Richter-Levin G, Klein E (2009) Physical stress differs from psychosocial stress in the pattern and time-course of behavioral responses, serum corticosterone and expression of plasticity-related genes in the rat. Stress 12:412–425

Kempes M, Matthys W, de Vries H, van Engeland H (2005) Reactive and proactive aggression in children–a review of theory, findings and the relevance for child and adolescent psychiatry. Eur Child Adolesc Psychiatry 14:11–19

Kempes M, de Vries H, Matthys W, van Engeland H, van Hooff J (2008) Differences in cortisol response affect the distinction of observed reactive and proactive aggression in children with aggressive behaviour disorders. J Neural Transm 115:139–147

Kessler MS, Murgatroyd C, Bunck M, Czibere L, Frank E, Jacob W, Horvath C, Muigg P, Holsboer F, Singewald N, Spengler D, Landgraf R (2007) Diabetes insipidus and, partially, low anxiety-related behaviour are linked to a SNP-associated vasopressin deficit in LAB mice. Eur J Neurosci 26:2857–2864

Keune PM, van der Heiden L, Várkuti B, Konicar L, Veit R, Birbaumer N (2012) Prefrontal brain asymmetry and aggression in imprisoned violent offenders. Neurosci Lett 515:191–195

Kew JN, Richards JG, Mutel V, Kemp JA (1998) Developmental changes in NMDA receptor glycine affinity and ifenprodil sensitivity reveal three distinct populations of NMDA receptors in individual rat cortical neurons. J Neurosci 18:1935–1943

Kiehl KA, Smith AM, Hare RD, Mendrek A, Forster BB, Brink J, Liddle PF (2001) Limbic abnormalities in affective processing by criminal psychopaths as revealed by functional magnetic resonance imaging. Biol Psychiatry 50:677–684

Kikusui T, Nakamura K, Kakuma Y, Mori Y (2006) Early weaning augments neuroendocrine stress responses in mice. Behav Brain Res 175:96–103

Kim U, Lee T (2012) Topography of descending projections from anterior insular and medial prefrontal regions to the lateral habenula of the epithalamus in the rat. Eur J Neurosci 35:1253–1269

Kim C, Jeon D, Kim YH, Lee CJ, Kim H, Shin HS (2009) Deletion of N-type Ca(2+) channel Ca(v)2.2 results in hyperaggressive behaviors in mice. J Biol Chem 284:2738–2745

Kim MJ, Loucks RA, Palmer AL, Brown AC, Solomon KM, Marchante AN, Whalen PJ (2011) The structural and functional connectivity of the amygdala: from normal emotion to pathological anxiety. Behav Brain Res 223:403–410

Kimbrell TA, George MS, Parekh PI, Ketter TA, Podell DM, Danielson AL, Repella JD, Benson BE, Willis MW, Herscovitch P, Post RM (1999) Regional brain activity during transient self-induced anxiety and anger in healthy adults. Biol Psychiatry 46:454–465

King JA, Blair RJ, Mitchell DG, Dolan RJ, Burgess N (2006) Doing the right thing: a common neural circuit for appropriate violent or compassionate behavior. Neuroimage 30:1069–1076

Kirillova GP, Vanyukov MM, Kirisci L, Reynolds M (2008) Physical maturation, peer environment, and the ontogenesis of substance use disorders. Psychiatry Res 158:43–53

Klein WP, Simon NG (1991) Timing of neonatal testosterone exposure in the differentiation of estrogenic regulatory systems for aggression. Physiol Behav 50:91–93

Klimes-Dougan B, Hastings PD, Granger DA, Usher BA, Zahn-Waxler C (2001) Adrenocortical activity in at-risk and normally developing adolescents: individual differences in salivary cortisol basal levels, diurnal variation, and responses to social challenges. Dev Psychopathol 13:695–719

Kling A, Lancaster J, Benitone J (1970) Amygdalectomy in the free-ranging vervet (Cercopithecus aethiops). J Psychiatr Res 7:191–199

Klotz U (2007) The role of pharmacogenetics in the metabolism of antiepileptic drugs: pharmacokinetic and therapeutic implications. Clin Pharmacokinet 46:271–279

Klüver H, Bucy PC (1938) An analysis of certain effects of bilateral temporal lobectomy in the Rhesus monkey, with special reference to psychic blindness. J Psychol 5:33–34, Cited based on Fountas and Smith, 2007

Knyshevski I, Connor DF, Harrison RJ, Ricci LA, Melloni RH Jr (2005) Persistent activation of select forebrain regions in aggressive, adolescent cocaine-treated hamsters. Behav Brain Res 159:277–286

Koch W, Schaaff N, Pöpperl G, Mulert C, Juckel G, Reicherzer M, Ehmer-von Geiso C, Möller HJ, Hegerl U, Tatsch K, Pogarell O (2007) [I-123] ADAM and SPECT in patients with borderline personality disorder and healthy control subjects. J Psychiatry Neurosci 32:234–240

Koenigsberg HW, Siever LJ, Lee H, Pizzarello S, New AS, Goodman M, Cheng H, Flory J, Prohovnik I (2009) Neural correlates of emotion processing in borderline personality disorder. Psychiatry Res 172:192–199

Kolb B (1974) Social behavior of rats with chronic prefrontal lesions. J Comp Physiol Psychol 87:466–474

Kollack-Walker S, Newman SW (1995) Mating and agonistic behavior produce different patterns of Fos immunolabeling in the male Syrian hamster brain. Neuroscience 66:721–736

Kollias SS, Alkadhi H, Jaermann T, Crelier G, Hepp-Reymond MC (2001) Identification of multiple nonprimary motor cortical areas with simple movements. Brain Res Brain Res Rev 36:185–195

Konoshenko MY, Timoshenko TV, Plyusnina IZ (2013) c-Fos activation and intermale aggression in rats selected for behavior toward humans. Behav Brain Res 237:103–106

Koob GF (2003) Neuroadaptive mechanisms of addiction: studies on the extended amygdala. Eur Neuropsychopharmacol 13:442–452

Koolhaas JM (1978) Hypothalamically induced intraspecific aggressive behaviour in the rat. Exp Brain Res 32:365–375

Koolhaas JM, De Boer SF, De Rutter AJ, Meerlo P, Sgoifo A (1997) Social stress in rats and mice. Acta Physiol Scand Suppl 640:69–72

Koolhaas JM, Bartolomucci A, Buwalda B, de Boer SF, Flügge G, Korte SM, Meerlo P, Murison R, Olivier B, Palanza P, Richter-Levin G, Sgoifo A, Steimer T, Stiedl O, van Dijk G, Wöhr M, Fuchs E (2011) Stress revisited: a critical evaluation of the stress concept. Neurosci Biobehav Rev 35:1291–1301

Krakowski M (2003) Violence and serotonin: influence of impulse control, affect regulation, and social functioning. J Neuropsychiatry Clin Neurosci 15:294–305

Kravitz EA, Huber R (2003) Aggression in invertebrates. Curr Opin Neurobiol 13:736–743

Kreuz LE, Rose RM (1972) Assessment of aggressive behaviour and plasma testosterone in a young criminal population. Psychosom Med 34:321–332

Kruesi MJ, Schmidt ME, Donnelly M, Hibbs ED, Hamburger SD (1989) Urinary free cortisol output and disruptive behavior in children. J Am Acad Child Adolesc Psychiatry 28:441–443

Kruk MR (1991) Ethology and pharmacology of hypothalamic aggression in the rat. Neurosci Biobehav Rev 15:527–538

Kruk MR, van der Poel AM, de Vos-Frerichs TP (1979) The induction of aggressive behaviour by electrical stimulation in the hypothalamus of male rats. Behaviour 70:292–322

Kruk MR, Van der Poel AM, Meelis W, Hermans J, Mostert PG, Mos J, Lohman AH (1983) Discriminant analysis of the localization of aggression-inducing electrode placements in the hypothalamus of male rats. Brain Res 260:61–79

Kruk MR, Westphal KG, Van Erp AM, van Asperen J, Cave BJ, Slater E, de Koning J, Haller J (1998) The hypothalamus: cross-roads of endocrine and behavioural regulation in grooming and aggression. Neurosci Biobehav Rev 23:163–177

Kruk MR, Halász J, Meelis W, Haller J (2004) Fast positive feedback between the adrenocortical stress response and a brain mechanism involved in aggressive behavior. Behav Neurosci 118:1062–1070

Kudielka BM, Wüst S (2010) Human models in acute and chronic stress: assessing determinants of individual hypothalamus-pituitary-adrenal axis activity and reactivity. Stress 13:1–14

Kuhlmann A, Bertsch K, Schmidinger I, Thomann P, Herpertz S (2013) Morphometric differences in central stress-regulating structures between women with and without borderline personality disorder. J Psychiat Neurosci 38:129–137

Kukekova AV, Johnson JL, Teiling C, Li L, Oskina IN, Kharlamova AV, Gulevich RG, Padte R, Dubreuil MM, Vladimirova AV, Shepeleva DV, Shikhevich SG, Sun Q, Ponnala L, Temnykh SV, Trut LN, Acland GM (2011) Sequence comparison of prefrontal cortical brain transcriptome from a tame and an aggressive silver fox (Vulpes vulpes). BMC Genomics 12:482

Laakso MP, Gunning-Dixon F, Vaurio O, Repo-Tiihonen E, Soininen H, Tiihonen J (2002) Prefrontal volumes in habitually violent subjects with antisocial personality disorder and type 2 alcoholism. Psychiatry Res 114:95–102

Laitinen LV (2001) Psychosurgery. Stereotact Funct Neurosurg 76:239–242

Lammers JH, Kruk MR, Meelis W, van der Poel AM (1988) Hypothalamic substrates for brain stimulation-induced attack, teeth-chattering and social grooming in the rat. Brain Res 449:311–327

Lammers JH, Meelis W, Kruk MR, van der Poel AM (1987) Hypothalamic substrates for brain stimulation-induced grooming, digging and circling in the rat. Brain Res 418:1–19

Landys MM, Goymann W, Schwabl I, Trapschuh M, Slagsvold T (2010) Impact of season and social challenge on testosterone and corticosterone levels in a year-round territorial bird. Horm Behav 58:317–325

Langhinrichsen-Rohling J (2005) Top 10 greatest "hits": important findings and future directions for intimate partner violence research. J Interpers Violence 20:108–118

Lanteaume L, Khalfa S, Régis J, Marquis P, Chauvel P, Bartolomei F (2007) Emotion induction after direct intracerebral stimulations of human amygdala. Cereb Cortex 17:1307–1313

Lapierre D, Braun CM, Hodgins S (1995) Ventral frontal deficits in psychopathy: neuropsychological test findings. Neuropsychologia 33:139–151

LeDoux J (2007) The amygdala. Curr Biol 17:R868–R874

Lee R, Coccaro E (2001) The neuropsychopharmacology of criminality and aggression. Can J Psychiatry 46:35–44

Lee G, Gammie SC (2007) GABA enhancement of maternal defense in mice: possible neural correlates. Pharmacol Biochem Behav 86:176–187

Lee G, Gammie SC (2010) GABAA receptor signaling in caudal periaqueductal gray regulates maternal aggression and maternal care in mice. Behav Brain Res 213:230–237

Lee GP, Bechara A, Adolphs R, Arena J, Meador KJ, Loring DW, Smith JR (1998) Clinical and physiological effects of stereotaxic bilateral amygdalotomy for intractable aggression. J Neuropsychiatry Clin Neurosci 10:413–420

Lee TM, Chan SC, Raine A (2008) Strong limbic and weak frontal activation to aggressive stimuli in spouse abusers. Mol Psychiatry 13:655–656

Lesch KP (2011) When the serotonin transporter gene meets adversity: the contribution of animal models to understanding epigenetic mechanisms in affective disorders and resilience. Curr Top Behav Neurosci 7:251–280

Lesch KP, Araragi N, Waider J, van den Hove D, Gutknecht L (2012) Targeting brain serotonin synthesis: insights into neurodevelopmental disorders with long-term outcomes related to negative emotionality, aggression and antisocial behaviour. Philos Trans R Soc Lond B Biol Sci 367:2426–2443

Leshner AI, Korn SJ, Mixon JF, Rosenthal C, Besser AK (1980) Effects of corticosterone on submissiveness in mice: some temporal and theoretical considerations. Physiol Behav 24:282–288

Levinson DM, Reeves DL, Buchanan DR (1980) Reductions in aggression and dominance status in guinea pigs following bilateral lesions in the basolateral amygdala or lateral septum. Physiol Behav 25:963–971

Li CS, Kosten TR, Sinha R (2006) Antisocial personality and stress-induced brain activation in cocaine-dependent patients. Neuroreport 17:243–247

Licata A, Taylor S, Berman M, Cranston J (1993) Effects of cocaine on human aggression. Pharmacol Biochem Behav 45:549–552

Lin D, Boyle MP, Dollar P, Lee H, Lein ES, Perona P, Anderson DJ (2011) Functional identification of an aggression locus in the mouse hypothalamus. Nature 470:221–226

Lindberg N, Tani P, Appelberg B, Naukkarinen H, Rimón R, Porkka-Heiskanen T, Virkkunen M (2003) Human impulsive aggression: a sleep research perspective. J Psychiatr Res 37:313–324

Lindenfors P, Tullberg BS (2011) Evolutionary aspects of aggression the importance of sexual selection. Adv Genet 75:7–22

Lipp HP (1978) Aggression and flight behaviour of the marmoset monkey Callithrix jacchus: an ethogram for brain stimulation studies. Brain Behav Evol 15:241–259

Lisciotto CA, DeBold JF, Miczek KA (1990) Sexual differentiation and the effects of alcohol on aggressive behavior in mice. Pharmacol Biochem Behav 35:357–362

Liu J, Portnoy J, Raine A (2012) Association between a marker for prenatal testosterone exposure and externalizing behavior problems in children. Dev Psychopathol 24:771–782

Loeber R, Hay D (1997) Key issues in the development of aggression and violence from childhood to early adulthood. Annu Rev Psychol 48:371–410

Loney BR, Butler MA, Lima EN, Counts CA, Eckel LA (2006) The relation between salivary cortisol, callous-unemotional traits, and conduct problems in an adolescent non-referred sample. J Child Psychol Psychiatry 47:30–36

López HS, Carrer HF (1985) Further studies on peripeduncular-hypothalamic pathways involved in sexual behavior in the female rat. Exp Neurol 88:241–252

Lopez NL, Vazquez DM, Olson SL (2004) An integrative approach to the neurophysiological substrates of social withdrawal and aggression. Dev Psychopathol 16:69–93

Lopez-Duran NL, Olson SL, Hajal NJ, Felt BT, Vazquez DM (2009) Hypothalamic pituitary adrenal axis functioning in reactive and proactive aggression in children. J Abnorm Child Psychol 37:169–182

Lorez K (1963) Das sogenannte Böse zur Naturgeschichte der Aggression. Dr. G Borotha-Schoeler, Vienna

Lotze M, Veit R, Anders S, Birbaumer N (2007) Evidence for a different role of the ventral and dorsal medial prefrontal cortex for social reactive aggression: an interactive fMRI study. Neuroimage 34:470–478

Lucion AB, De-Almeida RM, Da-Silva RS (1996) Territorial aggression, body weight, carbohydrate metabolism and testosterone levels of wild rats maintained in laboratory colonies. Braz J Med Biol Res 29:1657–1662

Lukas M, Bredewold R, Neumann ID, Veenema AH (2010) Maternal separation interferes with developmental changes in brain vasopressin and oxytocin receptor binding in male rats. Neuropharmacology 58:78–87

Luria AR (1980) Higher Cortical Functions in Man. Basic Books, New York, NY

Lynn SE (2008) Behavioral insensitivity to testosterone: why and how does testosterone alter paternal and aggressive behavior in some avian species but not others? Gen Comp Endocrinol 157:233–240

Lyons-Ruth K, Choi-Kain L, Pechtel P, Bertha E, Gunderson J (2011) Perceived parental protection and cortisol responses among young females with borderline personality disorder and controls. Psychiatry Res 189:426–432

Machado CJ, Bachevalier J (2006) The impact of selective amygdala, orbital frontal cortex, or hippocampal formation lesions on established social relationships in rhesus monkeys (Macaca mulatta). Behav Neurosci 120:761–786

Maeda H, Maki S (1986) Dopaminergic facilitation of recovery from amygdaloid lesions which affect hypothalamic defensive attack in cats. Brain Res 363:135–140

Mai JK, Asshcuer J, Paxinos G (1998) Atlas of the human brain. Academic, San Diego, CA

Mailly P, Aliane V, Groenewegen HJ, Haber SN, Deniau JM (2013) The rat prefrontostriatal system analyzed in 3D: evidence for multiple interacting functional units. J Neurosci 33:5718–5727

Mainardi M, Valenti G, Valsecchi P, Parmigiani S, Brain PF, Vescovi P (1987) Acute effects of low doses of two ACTH preparations on fighting behavior in male mice. Med Sci Res 15:649–650

Manharth A, Harris-Gerber L (2002) Surgical castration and the effect on aggression in rock hyrax (Procavia capensis). J Zoo Wildl Med 33:80–82

Maras A, Laucht M, Gerdes D, Wilhelm C, Lewicka S, Haack D, Malisova L, Schmidt MH (2003) Association of testosterone and dihydrotestosterone with externalizing behavior in adolescent boys and girls. Psychoneuroendocrinology 28:932–940

Marciello M, Sinnamon HM (1990) Locomotor stepping initiated by glutamate injections into the hypothalamus of the anesthetized rat. Behav Neurosci 104:980–990

Márquez C, Poirier GL, Cordero MI, Larsen MH, Groner A, Marquis J, Magistretti PJ, Trono D, Sandi C (2013) Peripuberty stress leads to abnormal aggression, altered amygdala and orbitofrontal reactivity and increased prefrontal MAOA gene expression. Transl Psychiatry 3:e216

Marsh AA, Finger EC, Mitchell DG, Reid ME, Sims C, Kosson DS, Towbin KE, Leibenluft E, Pine DS, Blair RJ (2008) Reduced amygdala response to fearful expressions in children and adolescents with callous-unemotional traits and disruptive behavior disorders. Am J Psychiatry 165:712–720

Marsman R, Swinkels SH, Rosmalen JG, Oldehinkel AJ, Ormel J, Buitelaar JK (2008) HPA-axis activity and externalizing behavior problems in early adolescents from the general population: the role of comorbidity and gender The TRAILS study. Psychoneuroendocrinology 33:789–798

Martin JR (1976) Motivated behaviors elicited from hypothalamus, midbrain, and pons of the guinea pig (Cavia porcellus). J Comp Physiol Psychol 90:1011–1034

Mathiak K, Weber R (2006) Toward brain correlates of natural behavior: fMRI during violent video games. Hum Brain Mapp 27:948–956

Matsuno-Yagi A, Mukohata Y (1977) Two possible roles of bacteriorhodopsin; a comparative study of strains of Halobacterium halobium differing in pigmentation. Biochem Biophys Res Commun 78:237–243

Matsuo K, Nicoletti M, Nemoto K, Hatch JP, Peluso MA, Nery FG, Soares JC (2009) A voxel-based morphometry study of frontal gray matter correlates of impulsivity. Hum Brain Mapp 30:1188–1195

Max JE, Levin HS, Schachar RJ, Landis J, Saunders AE, Ewing-Cobbs L, Chapman SB, Dennis M (2006) Predictors of personality change due to traumatic brain injury in children and adolescents six to twenty-four months after injury. J Neuropsychiatry Clin Neurosci 18:21–32

Maxson SC, Canastar A (2003) Conceptual and methodological issues in the genetics of mouse agonistic behavior. Horm Behav 44:258–262

Mazur A, Booth A (1998) Testosterone and dominance in men. Behav Brain Sci 21:353–363

McBurnett K, Lahey BB, Rathouz PJ, Loeber R (2000) Low salivary cortisol and persistent aggression in boys referred for disruptive behavior. Arch Gen Psychiatry 57:38–43

McBurnett K, King J, Scarpa A (2003) The HPA and the development of aggressive, antisocial, and substance abuse disorders. In: Cicchetti D, Walker E (eds) Neurodevelopmental mechanisms in psychopathology. Cambridge University Press, New York, NY, pp 324–344

McBurnett K, Raine A, Stouthamer-Loeber M, Loeber R, Kumar AM, Kumar M, Lahey BB (2005) Mood and hormone responses to psychological challenge in adolescent males with conduct problems. Biol Psychiatry 57:1109–1116

Mcdonald AJ, Mascagni F, Guo L (1996) Projections of the medial and lateral prefrontal cortices to the amygdala: a Phaseolus vulgaris leucoagglutinin study in the rat. Neuroscience 71:55–75

McDonald AJ, Shammah-Lagnado SJ, Shi C, Davis M (1999) Cortical afferents to the extended amygdala. Ann NY Acad Sci 877:309–338

McGregor A, Herbert J (1992) Differential effects of excitotoxic basolateral and corticomedial lesions of the amygdala on the behavioural and endocrine responses to either sexual or aggression-promoting stimuli in the male rat. Brain Res 574:9–20

Meaney MJ, Szyf M (2005) Environmental programming of stress responses through DNA methylation: life at the interface between a dynamic environment and a fixed genome. Dialogues Clin Neurosci 7:103–123

Meddle SL, Romero LM, Astheimer LB, Buttemer WA, Moore IT, Wingfield JC (2002) Steroid hormone interrelationships with territorial aggression in an Arctic-breeding songbird, Gambel's whitecrowned sparrow, Zonotrichia leucophrys gambelii. Horm Behav 42:212–221

Mehta PH, Beer J (2010) Neural mechanisms of the testosterone-aggression relation: the role of orbitofrontal cortex. J Cogn Neurosci 22:2357–2368

Mehta PH, Josephs RA (2010) Testosterone and cortisol jointly regulate dominance: evidence for a dual-hormone hypothesis. Horm Behav 58:898–906

Meletti S, Tassi L, Mai R, Fini N, Tassinari CA, Russo GL (2006) Emotions induced by intracerebral electrical stimulation of the temporal lobe. Epilepsia 47(Suppl 5):47–51

Melloni RH Jr, Ferris CF (1996) Adolescent anabolic steroid use and aggressive behavior in golden hamsters. Ann NY Acad Sci 794:372–375

Meloni EG, Reedy CL, Cohen BM, Carlezon WA Jr (2008) Activation of raphe efferents to the medial prefrontal cortex by corticotropin-releasing factor: correlation with anxiety-like behavior. Biol Psychiatry 63:832–839

Mempel E, Witkiewicz B, Stadnicki R, Luczywek E, Kucinski L, Pawlowski G et al (1980) The effect of medial amygdalotomy and anterior hippocampotomy on behavior and seizures in epileptic patients. Acta Neurochir Suppl 30:161–167

Mendelson J (1970) Self-induced drinking in rats: the qualitative identity of drive and reward systems in the lateral hypothalamus. Physiol Behav 5:925–930

Meunier M, Bachevalier J, Murray EA, Málková L, Mishkin M (1999) Effects of aspiration versus neurotoxic lesions of the amygdala on emotional responses in monkeys. Eur J Neurosci 11:4403–4418

Meyer JH, Wilson AA, Rusjan P, Clark M, Houle S, Woodside S, Arrowood J, Martin K, Colleton M (2008) Serotonin2A receptor binding potential in people with aggressive and violent behaviour. J Psychiatry Neurosci 33:499–508

Meyer-Lindenberg A, Buckholtz JW, Kolachana B, Hariri AR, Pezawas L, Blasi G, Wabnitz A, Honea R, Verchinski B, Callicott JH, Egan M, Mattay V, Weinberger DR (2006) Neural mechanisms of genetic risk for impulsivity and violence in humans. Proc Natl Acad Sci U S A 103:6269–6274

Miczek KA, Brykczynski T, Grossman SP (1974) Differential effects of lesions in the amygdala, periamygdaloid cortex, and stria terminalis on aggressive behaviors in rats. J Comp Physiol Psychol 87:760–771

Miczek KA, Weerts EM, Tornatzky W, DeBold JF, Vatne TM (1992) Alcohol and "bursts" of aggressive behavior: ethological analysis of individual differences in rats. Psychopharmacology (Berl) 107:551–563

Miczek KA, DeBold JF, van Erp AM (1994) Neuropharmacological characteristics of individual differences in alcohol effects on aggression in rodents and primates. Behav Pharmacol 5:407–421

Miczek KA, Fish EW, De Bold JF, De Almeida RM (2002) Social and neural determinants of aggressive behavior: pharmacotherapeutic targets at serotonin, dopamine and gamma-aminobutyric acid systems. Psychopharmacology (Berl) 163:434–458

Miczek KA, de Boer SF, Haller J (2013) Excessive aggression as model of violence: a critical evaluation of current preclinical methods. Psychopharmacology (Berl) 226:445–458

Mikics E, Kruk MR, Haller J (2004) Genomic and non-genomic effects of glucocorticoids on aggressive behavior in male rats. Psychoneuroendocrinology 29:618–635

Mikics E, Barsy B, Haller J (2007) The effect glucocorticoids on aggressiveness in established colonies of rats. Psychoneuroendocrinology 32:160–170

Miles DR, Carey G (1997) Genetic and environmental architecture of human aggression. J Pers Soc Psychol 72:207–217

Millard A, Gentsch C (2006) Competition for sucrose pellets in tetrads of male Wistar, Fischer or Sprague–Dawley rats: is intra-group ranking reflected in the level of anxiety? Behav Brain Res 168:243–254
Millesi E, Hoffmann IE, Steurer S, Metwaly M, Dittami JP (2002) Vernal changes in the behavioral and endocrine responses to GnRH application in male European ground squirrels. Horm Behav 41:51–58
Millet K, Dewitte S (2009) The presence of aggression cues inverts the relation between digit ratio (2D:4D) and prosocial behaviour in a dictator game. Br J Psychol 100:151–162
Miner LL, Elmer GI, Pieper JO, Marley RJ (1993) Aggression modulates genetic influences on morphine analgesia as assessed using a classical mendelian cross analysis. Psychopharmacology (Berl) 111:17–22
Minzenberg MJ, Fan J, New AS, Tang CY, Siever LJ (2007) Fronto-limbic dysfunction in response to facial emotion in borderline personality disorder: an event-related fMRI study. Psychiatry Res 155:231–243
Moldow RL, Fischman AJ (1987) Cocaine induced secretion of ACTH, beta-endorphin, and corticosterone. Peptides 8:819–822
Money J (1965) Influence of hormones on sexual behavior. Annu Rev Med 16:67–82
Mong JA, Pfaff DW (2003) Hormonal and genetic influences underlying arousal as it drives sex and aggression in animal and human brains. Neurobiol Aging 24(Suppl 1):S83–S88
Montag C, Weber B, Trautner P, Newport B, Markett S, Walter NT, Felten A, Reuter M (2012) Does excessive play of violent first-person-shooter-video-games dampen brain activity in response to emotional stimuli? Biol Psychol 89:107–111
Moore MC, Marler CA (1987) Effects of testosterone manipulations on nonbreeding season territorial aggression in free-living male lizards, Sceloporus jarrovi. Gen Comp Endocrinol 65:225–232
Morandotti N, Dima D, Jogia J, Frangou S, Sala M, Vidovich GZ, Lazzaretti M, Gambini F, Marraffini E, d'Allio G, Barale F, Zappoli F, Caverzasi E, Brambilla P (2013) Childhood abuse is associated with structural impairment in the ventrolateral prefrontal cortex and aggressiveness in patients with borderline personality disorder. Psychiatry Res 213:18–23
Mos J, Kruk MR, Van der Poel AM, Meelis W (1982) Aggressive behavior induced by electrical stimulation in the midbrain central gray of male rats. Aggress Behav 8:261–284
Mos J, Lammers JHCM, Van der Poel AM, Bermond B, Meelis W, Kruk MR (1983) Effects of midbrain central gray lesions on spontaneous and electrically induced aggression in the rat. Aggress Behav 9:133–155
Moss HB, Vanyukov MM, Martin CS (1995) Salivary cortisol responses and the risk for substance abuse in prepubertal boys. Biol Psychiatry 38:547–555
Motelica-Heino I, Edwards DA, Roffi J (1993) Intermale aggression in mice: does hour of castration after birth influence adult behaviour? Physiol Behav 53:1017–1019
Mpakopoulou M, Gatos H, Brotis A, Paterakis KN, Fountas KN (2008) Stereotactic amygdalotomy in the management of severe aggressive behavioral disorders. Neurosurg Focus 25:E6
Mukai M, Replogle K, Drnevich J, Wang G, Wacker D, Band M, Clayton DF, Wingfield JC (2009) Seasonal differences of gene expression profiles in song sparrow (Melospiza melodia) hypothalamus in relation to territorial aggression. PLoS One 4:e8182
Muller MN, Wrangham RW (2004) Dominance, aggression and testosterone in wild chimpanzees: a test of the 'challenge' hypothesis. Anim Behav 67:113–123
Müller JL, Sommer M, Wagner V, Lange K, Taschler H, Röder CH, Schuierer G, Klein HE, Hajak G (2003) Abnormalities in emotion processing within cortical and subcortical regions in criminal psychopaths: evidence from a functional magnetic resonance imaging study using pictures with emotional content. Biol Psychiatry 54:152–162
Munro AD, Pitcher TJ (1985) Steroid hormones and agonistic behavior in a cichlid teleost, Aequidens pulcher. Horm Behav 19:353–371

Murray-Close D, Han G, Cicchetti D, Crick NR, Rogosch FA (2008) Neuroendocrine regulation and physical and relational aggression: the moderating roles of child maltreatment and gender. Dev Psychol 44:1160–1176
Myers B, Mark Dolgas C, Kasckow J, Cullinan WE, Herman JP (2013) Central stress-integrative circuits: forebrain glutamatergic and GABAergic projections to the dorsomedial hypothalamus, medial preoptic area, and bed nucleus of the stria terminalis. Brain Struct Funct [Epub ahead of print]
Nagel G, Ollig D, Fuhrmann M, Kateriya S, Musti AM, Bamberg E, Hegemann P (2002) Channelrhodopsin-1: a light-gated proton channel in green algae. Science 296:2395–2398
Narabayashi H, Uno M (1966) Long range results of stereotaxic amygdalotomy for behavior disorders. Confin Neurol 27:168–171
Narabayashi H, Nagao T, Saito Y, Yoshida M, Nagahata M (1963) Stereotaxic amygdalotomy for behavior disorders. Arch Neurol 9:1–16, Cited based on Fountas and Smith, 2007
Natarajan D, Caramaschi D (2010) Animal violence demystified. Front Behav Neurosci 4:9
Natarajan D, de Boer SF, Koolhaas JM (2009a) Lack of differential serotonin biosynthesis capacity in genetically selected low and high aggressive mice. Physiol Behav 98:411–415
Natarajan D, de Vries H, de Boer SF, Koolhaas JM (2009b) Violent phenotype in SAL mice is inflexible and fixed in adulthood. Aggress Behav 35:430–436
Natarajan D, de Vries H, Saaltink DJ, de Boer SF, Koolhaas JM (2009c) Delineation of violence from functional aggression in mice: an ethological approach. Behav Genet 39:73–90
Neddens J, Bagorda F, Busche A, Horstmann S, Moll GH, Dawirs RR, Teuchert-Noodt G (2003) Epigenetic factors differentially influence postnatal maturation of serotonin (5-HT) innervation in cerebral cortex of gerbils: interaction of rearing conditions and early methamphetamine challenge. Brain Res Dev Brain Res 146:119–130
Nelson RJ (2005) Effects of nitric oxide on the HPA axis and aggression. Novartis Found Symp 268:147–160
Nelson RJ, Chiavegatto S (2001) Molecular basis of aggression. Trends Neurosci 24:713–719
Nelson RJ, Trainor BC, Chiavegatto S, Demas GE (2006) Pleiotropic contributions of nitric oxide to aggressive behavior. Neurosci Biobehav Rev 30:346–355
Nephew BC, Kahn SA, Romero LM (2003) Heart rate and behavior are regulated independently of corticosterone following diverse acute stressors. Gen Comp Endocrinol 133:173–180
Nephew BC, Caffrey MK, Felix-Ortiz AC, Ferris CF, Febo M (2009) Blood oxygen level-dependent signal responses in corticolimbic 'emotions' circuitry of lactating rats facing intruder threat to pups. Eur J Neurosci 30:934–945
Nesher E, Gross M, Lisson S, Tikhonov T, Yadid G, Pinhasov A (2013) Differential responses to distinct psychotropic agents of selectively bred dominant and submissive animals. Behav Brain Res 236:225–235
Neumann ID (2001) Alterations in behavioral and neuroendocrine stress coping strategies in pregnant, parturient and lactating rats. Prog Brain Res 133:143–152
Neumann ID, Toschi N, Ohl F, Torner L, Kromer SA (2001) Maternal defence as an emotional stressor in female rats: correlation of neuroendocrine and behavioural parameters and involvement of brain oxytocin. Eur J Neurosci 13:1016–1024
Neumann ID, Veenema AH, Beiderbeck DI (2010) Aggression and anxiety: social context and neurobiological links. Front Behav Neurosci 4:12
New AS, Hazlett EA, Buchsbaum MS, Goodman M, Reynolds D, Mitropoulou V, Sprung L, Shaw RB Jr, Koenigsberg H, Platholi J, Silverman J, Siever LJ (2002) Blunted prefrontal cortical 18fluorodeoxyglucose positron emission tomography response to meta-chlorophenylpiperazine in impulsive aggression. Arch Gen Psychiatry 59:621–629
New AS, Hazlett EA, Buchsbaum MS, Goodman M, Mitelman SA, Newmark R, Trisdorfer R, Haznedar MM, Koenigsberg HW, Flory J, Siever LJ (2007) Amygdala-prefrontal disconnection in borderline personality disorder. Neuropsychopharmacology 32:1629–1640
New AS, Hazlett EA, Newmark RE, Zhang J, Triebwasser J, Meyerson D, Lazarus S, Trisdorfer R, Goldstein KE, Goodman M, Koenigsberg HW, Flory JD, Siever LJ, Buchsbaum MS (2009)

Laboratory induced aggression: a positron emission tomography study of aggressive individuals with borderline personality disorder. Biol Psychiatry 66:1107–1114
Newman SW (1999) The medial extended amygdala in male reproductive behavior. A node in the mammalian social behavior network. Ann NY Acad Sci 877:242–257
Nicolas LB, Pinoteau W, Papot S, Routier S, Guillaumet G, Mortaud S (2001) Aggressive behavior induced by the steroid sulfatase inhibitor COUMATE and by DHEAS in CBA/H mice. Brain Res 922:216–222
Niedermeyer E (1998) Frontal lobe functions and dysfunctions. Clin Electroencephalogr 29:79–90
Nieh EH, Kim SY, Namburi P, Tye KM (2013) Optogenetic dissection of neural circuits underlying emotional valence and motivated behaviors. Brain Res 1511:73–92
Noblett KL, Coccaro EF (2005) Molecular genetics of personality. Curr Psychiatry Rep 7:73–80
Nyby JG (2008) Reflexive testosterone release: a model system for studying the nongenomic effects of testosterone upon male behavior. Front Neuroendocrinol 29:199–210
Oakes ME, Coover GD (1997) Effects of small amygdala lesions on fear, but not aggression, in the rat. Physiol Behav 61:45–55
O'Connor DB, Archer J, Hair WM, Wu FC (2002) Exogenous testosterone, aggression, and mood in eugonadal and hypogonadal men. Physiol Behav 75:557–566
Oesterhelt D, Stoeckenius W (1971) Rhodopsin-like protein from the purple membrane of Halobacterium halobium. Nat New Biol 233:149–152
Olah S, Fule M, Komlosi G, Varga C, Baldi R, Barzo P, Tamas G (2009) Regulation of cortical microcircuits by unitary GABA-mediated volume transmission. Nature 461:1278–1281
Olivier B (2004) Serotonin and aggression. Ann NY Acad Sci 1036:382–392
O'Neal CR, Brotman LM, Huang KY, Gouley KK, Kamboukos D, Calzada EJ, Pine DS (2010) Understanding relations among early family environment, cortisol response, and child aggression via a prevention experiment. Child Dev 81:290–305
Oosterlaan J, Geurts HM, Knol DL, Sergeant JA (2005) Low basal salivary cortisol is associated with teacher-reported symptoms of conduct disorder. Psychiatry Res 134:1–10
Oquendo MA, Russo SA, Underwood MD, Kassir SA, Ellis SP, Mann JJ, Arango V (2006) Higher postmortem prefrontal 5-HT2A receptor binding correlates with lifetime aggression in suicide. Biol Psychiatry 59:235–243
Orpinas P, Frankowski R (2001) A self-report measure of aggressive behavior for young adolescents. J Early Adolescence 21:50–67
Ortiz R, Armario A, Castellanos JM (1984) Post-weaning differential housing and testosterone secretion in male mice. Experientia 40:1428–1429
Osumi T, Nakao T, Kasuya Y, Shinoda J, Yamada J, Ohira H (2012) Amygdala dysfunction attenuates frustration-induced aggression in psychopathic individuals in a non-criminal population. J Affect Disord 142:331–338
Overli O, Kotzian S, Winberg S (2002) Effects of cortisol on aggression and locomotor activity in rainbow trout. Horm Behav 42:53–61
Overli O, Korzan WJ, Larson ET, Winberg S, Lepage O, Pottinger TG, Renner KJ, Summers CH (2004) Behavioral and neuroendocrine correlates of displaced aggression in trout. Horm Behav 45:324–349
Oyegbile TO, Marler CA (2005) Winning fights elevates testosterone levels in California mice and enhances future ability to win fights. Horm Behav 48:259–267
Oyegbile TO, Marler CA (2006) Weak winner effect in a less aggressive mammal: correlations with corticosterone but not testosterone. Physiol Behav 89:171–179
Paine TA, Asinof SK, Diehl GW, Frackman A, Leffler J (2013) Medial prefrontal cortex lesions impair decision-making on a rodent gambling task: reversal by D1 receptor antagonist administration. Behav Brain Res 243:247–254
Painuly NP, Grover S, Gupta N, Mattoo SK (2011) Prevalence of anger attacks in depressive and anxiety disorders: implications for their construct? Psychiatry Clin Neurosci 65:165–174
Pajer K, Gardner W, Rubin RT, Perel J, Neal S (2001) Decreased cortisol levels in adolescent girls with conduct disorder. Arch Gen Psychiatry 58:297–302

Pajer K, Tabbah R, Gardner W, Rubin RT, Czambel RK, Wang Y (2006) Adrenal androgen and gonadal hormone levels in adolescent girls with conduct disorder. Psychoneuroendocrinology 31:1245–1256

Pan Y, Xu L, Young KA, Wang Z, Zhang Z (2010) Agonistic encounters and brain activation in dominant and subordinate male greater long-tailed hamsters. Horm Behav 58:478–484

Panksepp J (2011) The basic emotional circuits of mammalian brains: do animals have affective lives? Neurosci Biobehav Rev 35:1791–1804

Paperny DM, Deisher RW (1983) Maltreatment of adolescents: the relationship to a predisposition toward violent behavior and delinquency. Adolescence 18:499–506

Paquette J, Underwood M (1999) Gender differences in young adolescents' experiences of peer victimization; social and physical aggression. Merrill-Palmer Quart 45:242–266

Pardini DA, Phillips M (2010) Neural responses to emotional and neutral facial expressions in chronically violent men. J Psychiatry Neurosci 35:390–398

Pardini M, Krueger F, Hodgkinson C, Raymont V, Ferrier C, Goldman D, Strenziok M, Guida S, Grafman J (2011) Prefrontal cortex lesions and MAO-A modulate aggression in penetrating traumatic brain injury. Neurology 76:1038–1045

Pare D, Duvarci S (2012) Amygdala microcircuits mediating fear expression and extinction. Curr Opin Neurobiol 22:717–723

Parmigiani S, Brain PF, Mainardi D, Brunoni V (1988) Different patterns of biting attack employed by lactating female mice (Musdomesticus) in encounters with male and female conspecific intruders. J Comp Psychol 102:287–293

Parmigiani S, Dadomo H, Bartolomucci A, Brain PF, Carbucicchio A, Costantino C, Ferrari PF, Palanza P, Volpi R (2009) Personality traits and endocrine response as possible asymmetry factors of agonistic outcome in karate athletes. Aggress Behav 35:324–333

Parsey RV, Oquendo MA, Simpson NR, Ogden RT, Van Heertum R, Arango V, Mann JJ (2002) Effects of sex, age, and aggressive traits in man on brain serotonin 5-HT1A receptor binding potential measured by PET using [C-11]WAY-100635. Brain Res 954:173–182

Partecke J, Schwabl H (2008) Organizational effects of maternal testosterone on reproductive behavior of adult house sparrows. Dev Neurobiol 68:1538–1548

Passamonti L, Fera F, Magariello A, Cerasa A, Gioia MC, Muglia M, Nicoletti G, Gallo O, Provinciali L, Quattrone A (2006) Monoamine oxidase-a genetic variations influence brain activity associated with inhibitory control: new insight into the neural correlates of impulsivity. Biol Psychiatry 59:334–340

Passamonti L, Crockett MJ, Apergis-Schoute AM, Clark L, Rowe JB, Calder AJ, Robbins TW (2012) Effects of acute tryptophan depletion on prefrontal-amygdala connectivity while viewing facial signals of aggression. Biol Psychiatry 71:36–43

Pasterski V, Hindmarsh P, Geffner M, Brook C, Brain C, Hines M (2007) Increased aggression and activity level in 3- to 11-year-old girls with congenital adrenal hyperplasia (CAH). Horm Behav 52:368–374

Patel BD, Barzman DH (2013) Pharmacology and pharmacogenetics of pediatric ADHD with associated aggression: a review. Psychiatr Q 84:407–415

Pavlov KA, Chistiakov DA, Chekhonin VP (2012) Genetic determinants of aggression and impulsivity in humans. J Appl Genet 53:61–82

Paxinos G, Watson C (1998) The rat brain in stereotaxic coordinates. Academic

Payer DE, Lieberman MD, London ED (2011) Neural correlates of affect processing and aggression in methamphetamine dependence. Arch Gen Psychiatry 68:271–282

Payer DE, Nurmi EL, Wilson SA, McCracken JT, London ED (2012) Effects of methamphetamine abuse and serotonin transporter gene variants on aggression and emotion-processing neurocircuitry. Transl Psychiatry 2:e80

Pedersen WC, Gonzales C, Miller N (2000) The moderating effect of trivial triggering provocation on displaced aggression. J Pers Soc Psychol 78:913–927

Pedrosa-Sanchez M, Sola RG (2003) Modern day psychosurgery: a new approach to neurosurgery in psychiatric disease. Rev Neurol 36:887–897

Perroud N, Salzmann A, Prada P, Nicastro R, Hoeppli ME, Furrer S, Ardu S, Krejci I, Karege F, Malafosse A (2013) Response to psychotherapy in borderline personality disorder and methylation status of the BDNF gene. Transl Psychiatry 3:e207

Peters PJ, Bronson FH, Whitsett JM (1972) Neonatal castration and intermale aggression in mice. Physiol Behav 8:265–268

Petrides M (2005) Lateral prefrontal cortex: architectonic and functional organization. Philos Trans R Soc Lond B Biol Sci 360:781–795

Picard N, Strick PL (2001) Imaging the premotor areas. Curr Opin Neurobiol 11:663–672

Plusquellec P, Bouissou M (2001) Behavioural characteristics of two dairy breeds of cows selected (Hérens) or not (Brune des Alpes) for fighting and dominance ability. Appl Anim Behav Sci 72:1–21

Plyusnina I, Oskina I (1997) Behavioral and adrenocortical responses to open-field test in rats selected for reduced aggressiveness toward humans. Physiol Behav 61:381–385

Poggioli R, Vergoni AV, Santi R, Carani C, Baraghini GF, Zini D, Marrama P, Bertolini A (1984) Sexual behavior of male rats: influence of short- and long-term adrenalectomy. Horm Behav 18:79–85

Popma A, Jansen LM, Vermeiren R, Steiner H, Raine A, Van Goozen SH, van Engeland H, Doreleijers TA (2006) Hypothalamus pituitary adrenal axis and autonomic activity during stress in delinquent male adolescents and controls. Psychoneuroendocrinology 31:948–957

Popma A, Doreleijers TA, Jansen LM, Van Goozen SH, Van Engeland H, Vermeiren R (2007a) The diurnal cortisol cycle in delinquent male adolescents and normal controls. Neuropsychopharmacology 32:1622–1628

Popma A, Vermeiren R, Geluk CA, Rinne T, van den Brink W, Knol DL, Jansen LM, van Engeland H, Doreleijers TA (2007b) Cortisol moderates the relationship between testosterone and aggression in delinquent male adolescents. Biol Psychiatry 61:405–411

Potegal M (1991) Attack priming and satiation in female golden hamsters: tests of some alternatives to the aggression arousal interpretation. Aggress Behav 17:327–335

Potegal M (2012) Temporal and frontal lobe initiation and regulation of the top-down escalation of anger and aggression. Behav Brain Res 231:386–395

Potegal M, Huhman K, Moore T, Meyerhoff J (1993) Conditioned defeat in the Syrian golden hamster (Mesocricetus auratus). Behav Neural Biol 60:93–102

Potegal M, Ferris CF, Hebert M, Meyerhoff J, Skaredoff L (1996a) Attack priming in female Syrian golden hamsters is associated with a c-fos-coupled process within the corticomedial amygdala. Neuroscience 75:869–880

Potegal M, Hebert M, DeCoster M, Meyerhoff JL (1996b) Brief, high-frequency stimulation of the corticomedial amygdala induces a delayed and prolonged increase of aggressiveness in male Syrian golden hamsters. Behav Neurosci 110:401–412

Pott CB, Kramer SZ, Siegel A (1987) Central gray modulation of affective defense is differentially sensitive to naloxone. Physiol Behav 40:207–213

Potts GF, George MR, Martin LE, Barratt ES (2006) Reduced punishment sensitivity in neural systems of behavior monitoring in impulsive individuals. Neurosci Lett 397:130–134

Poustka L, Maras A, Hohm E, Fellinger J, Holtmann M, Banaschewski T, Lewicka S, Schmidt MH, Esser G, Laucht M (2010) Negative association between plasma cortisol levels and aggression in a high-risk community sample of adolescents. J Neural Transm 117:621–627

Powell DA, Francis J, Braman MJ, Schneiderman N (1969) Frequency of attack in shock-elicited aggression as a function of the performance of individual rats. J Exp Anal Behav 12:817–823

Prinssen EP, Nicolas LB, Klein S, Grundschober C, Lopez-Lopez C, Kessler MS, Bruns A, von Kienlin M, Wettstein JG, Moreau JL, Risterucci C (2012) Imaging trait anxiety in high anxiety F344 rats: focus on the dorsomedial prefrontal cortex. Eur Neuropsychopharmacol 22:441–451

Putkonen PT (1966) Attack elicited by forebrain and hypothalamic stimulation in the chicken. Experientia 22:405–407

Quadagno DM, Briscoe R, Quadagno JS (1977) Effect of perinatal gonadal hormones on selected nonsexual behavior patterns: a critical assessment of the human and nonhuman literature. Psychol Bull 84:62–80
Quadagno DM, McQuitty C, McKee J, Koelliker L, Wolfe G, Johnson DC (1987) The effects of intrauterine position on competition and behavior in the mouse. Physiol Behav 41:639–642
Raine A (2002) Annotation: the role of prefrontal deficits, low autonomic arousal, and early health factors in the development of antisocial and aggressive behavior in children. J Child Psychol Psychiatry 43:417–434
Raine A, Yang Y (2006) Neural foundations to moral reasoning and antisocial behavior. Soc Cogn Affect Neurosci 1:203–213
Raine A, Buchsbaum MS, Stanley J (1992) Selective reductions in prefrontal glucose metabolism in murderers assessed with positron emission tomography. Psychophysiology 29(suppl 4A):58
Raine A, Buchsbaum MS, Stanley J, Lottenberg S, Abel L, Stoddard J (1994) Selective reductions in prefrontal glucose metabolism in murderers. Biol Psychiat 36:365–373
Raine A, Buchsbaum M, LaCasse L (1997) Brain abnormalities in murderers indicated by positron emission tomography. Biol Psychiatry 42:495–508
Raine A, Lencz T, Bihrle S, LaCasse L, Colletti P (2000) Reduced prefrontal gray matter volume and reduced autonomic activity in antisocial personality disorder. Arch Gen Psychiatry 57:119–127, discussion 128–129
Raine A, Dodge K, Loeber R, Gatzke-Kopp L, Lynam D, Reynolds C, Stouthamer-Loeber M, Liu J (2006) The reactive-proactive aggression questionnaire: differential correlates of reactive and proactive aggression in adolescent boys. Aggress Behav 32:159–171
Raine A, Yang Y, Narr KL, Toga AW (2011) Sex differences in orbitofrontal gray as a partial explanation for sex differences in antisocial personality. Mol Psychiatry 16:227–236
Ramamurthi B (1988) Stereotactic operation in behaviour disorders. Amygdalotomy and hypothalamotomy. Acta Neurochir Suppl 44:152–157
Ramenofsky M (1985) Acute changes in plasma steroids and agonistic behavior in male Japanese quail. Gen Comp Endocrinol 60:116–128
Ranck JB Jr (1975) Which elements are excited in electrical stimulation of mammalian central nervous system: a review. Brain Res 98:417–440
Rapoport A, Chammah A (1966) The game of chicken. Am Behav Scientist 10:10–28
Rasanen P, Hakko H, Visuri S, Paanila J, Kapanen P, Suomela T, Tiihonen J (1999) Serum testosterone levels, mental disorders and criminal behavior. Acta Psychiat Scand 99:348–352
Rasia-Filho AA, Londero RG, Achaval M (2000) Functional activities of the amygdala: an overview. J Psychiatry Neurosci 25:14–23
Ray JP, Price JL (1992) The organization of the thalamocortical connections of the mediodorsal thalamic nucleus in the rat, related to the ventral forebrain-prefrontal cortex topography. J Comp Neurol 323:167–197
Razin A, Riggs AD (1980) DNA methylation and gene function. Science 210:604–610
Reef J, Diamantopoulou S, van Meurs I, Verhulst FC, van der Ende J (2011) Developmental trajectories of child to adolescent externalizing behavior and adult DSM-IV disorder: results of a 24-year longitudinal study. Soc Psychiatry Psychiatr Epidemiol 46:1233–1241
Regenbogen C, Herrmann M, Fehr T (2010) The neural processing of voluntary completed, real and virtual violent and nonviolent computer game scenarios displaying predefined actions in gamers and nongamers. Soc Neurosci 5:221–240
Reijntjes A, Dekovic M, Telch MJ (2007) Support for the predictive validity of the SASC–R: linkages with reactions to an in vivo peer evaluation manipulation. J Anxiety Disord 21:903–917
Reinisch JM (1981) Prenatal exposure to synthetic progestins increases potential for aggression in humans. Science 211:1171–1173
Reinisch JM, Sanders SA (1986) A test of sex differences in aggressive response to hypothetical conflict situations. J Pers Soc Psychol 50:1045–1049

Rejeski WJ, Gregg E, Kaplan JR, Manuck SB (1990) Anabolic-androgenic steroids: effects on social behavior and baseline heart rate. Health Psychol 9:774–791

Rempel-Cclower NL, Barbas H (1998) Topographic organization of connections between the hypothalamus and prefrontal cortex in the rhesus monkey. Comp Neurol 398:393–419

Reul JM, Chandramohan Y (2007) Epigenetic mechanisms in stress-related memory formation. Psychoneuroendocrinology 32(Suppl 1):S21–S25

Reul JM, Labeur MS, Grigoriadis DE, De Souza EB, Holsboer F (1994) Hypothalamic-pituitary-adrenocortical axis changes in the rat after long-term treatment with the reversible monoamine oxidase-A inhibitor moclobemide. Neuroendocrinology 60:509–519

Ricci LA, Grimes JM, Knyshevski I, Melloni RH Jr (2005) Repeated cocaine exposure during adolescence alters glutamic acid decarboxylase-65 (GAD65) immunoreactivity in hamster brain: correlation with offensive aggression. Brain Res 1035:131–138

Ricci LA, Rasakham K, Grimes JM, Melloni RH Jr (2006) Serotonin-1A receptor activity and expression modulate adolescent anabolic/androgenic steroid-induced aggression in hamsters. Pharmacol Biochem Behav 85:1–11

Ricci LA, Grimes JM, Melloni RH Jr (2007) Lasting changes in neuronal activation patterns in select forebrain regions of aggressive, adolescent anabolic/androgenic steroid-treated hamsters. Behav Brain Res 176:344–352

Ricci LA, Schwartzer JJ, Melloni RH Jr (2009) Alterations in the anterior hypothalamic dopamine system in aggressive adolescent AAS-treated hamsters. Horm Behav 55:348–355

Riggs AD (1975) X inactivation, differentiation, and DNA methylation. Cytogenet Cell Genet 14:9–25

Rilling JK, Winslow JT, Kilts CD (2004) The neural correlates of mate competition in dominant male rhesus macaques. Biol Psychiatry 56:364–375

Rilling JK, Glenn AL, Jairam MR, Pagnoni G, Goldsmith DR, Elfenbein HA, Lilienfeld SO (2007) Neural correlates of social cooperation and non-cooperation as a function of psychopathy. Biol Psychiatry 61:1260–1271

Roberts WW, Steinberg ML, Means LW (1967) Hypothalamic mechanisms for sexual, aggressive, and other motivational behaviors in the opossium, Didelphis virginiana. J Comp Physiol Psychol 64:1–15

Roeling TAP, Veening JG, Kruk MR, Nieuwenhuys R (1990) Grooming behaviour elicited by kainic acid evoked cell body stimulation in the hypothalamus of the rat. Neurosci Res Commun 6:111–118

Roeling TA, Kruk MR, Schuurmans R, Veening JG (1993) Behavioural responses of bicucculline methiodide injections into the ventral hypothalamus of freely moving, socially interacting rats. Brain Res 615:121–127

Roeling TA, Veening JG, Kruk MR, Peters JP, Vermelis ME, Nieuwenhuys R (1994) Efferent connections of the hypothalamic "aggression area" in the rat. Neuroscience 59:1001–1024

Roepke S, Ziegenhorn A, Kronsbein J, Merkl A, Bahri S, Lange J, Lübbert H, Schweiger U, Heuser I, Lammers CH (2010) Incidence of polycystic ovaries and androgen serum levels in women with borderline personality disorder. J Psychiatr Res 44:847–852

Romaniuk A (1965) Representation of aggression and flight reactions in the hypothalamus of the cat. Acta Biol Exp 25:177–186

Romero-Martínez A, Lila M, Sariñana-González P, González-Bono E, Moya-Albiol L (2013) High testosterone levels and sensitivity to acute stress in perpetrators of domestic violence with low cognitive flexibility and impairments in their emotional decoding process: a preliminary study. Aggress Behav 39:355–369

Ros AF, Dieleman SJ, Groothuis TG (2002) Social stimuli, testosterone, and aggression in gull chicks: support for the challenge hypothesis. Horm Behav 41:334–342

Rose RM, Holaday JW, Bernstein IS (1971) Plasma testosterone, dominance rank and aggressive behaviour in male rhesus monkeys. Nature 231:366–368

Rosell DR, Thompson JL, Slifstein M, Xu X, Frankle WG, New AS, Goodman M, Weinstein SR, Laruelle M, Abi-Dargham A, Siever LJ (2010) Increased serotonin 2A receptor availability in

the orbitofrontal cortex of physically aggressive personality disordered patients. Biol Psychiatry 67:1154–1162
Rosenzweig S, Ludwig DJ, Adelman S (1975) Retest reliability of the Rosenzweig Picture-Frustration Study and similar semiprojective techniques. J Pers Assess 39:3–12
Rouwette T, Vanelderen P, Roubos EW, Kozicz T, Vissers K (2012) The amygdala, a relay station for switching on and off pain. Eur J Pain 16:782–792
Roy AK, Shehzad Z, Margulies DS, Kelly AM, Uddin LQ, Gotimer K, Biswal BB, Castellanos FX, Milham MP (2009) Functional connectivity of the human amygdala using resting state fMRI. Neuroimage 45:614–626
Rubin RT, Reimsch JM, Haskett RF (1981) Postnatal gonadal steroid effects on human behavior. Science 211:1318–1324
Rubinstein LJ (1976) Current concepts in neuro-oncology. Adv Neurol 15:1–25
Rudebeck PH, Walton ME, Millette BH, Shirley E, Rushworth MF, Bannerman DM (2007) Distinct contributions of frontal areas to emotion and social behaviour in the rat. Eur J Neurosci 26:2315–2326
Ruiz-de-la-torre JL, Manteca X (1999) Effects of testosterone on aggressive behaviour after social mixing in male lambs. Physiol Behav 68:109–113
Rüsch N, van Elst LT, Ludaescher P, Wilke M, Huppertz HJ, Thiel T, Schmahl C, Bohus M, Lieb K, Hesslinger B, Hennig J, Ebert D (2003) A voxel-based morphometric MRI study in female patients with borderline personality disorder. Neuroimage 20:385–392
Rushworth MF, Walton ME, Kennerley SW, Bannerman DM (2004) Action sets and decisions in the medial frontal cortex. Trends Cogn Sci 8:410–417
Rutter M, Moffitt TE, Caspi A (2006) Gene-environment interplay and psychopathology: multiple varieties but real effects. J Child Psychol Psychiat 47:226–261
Ruttle PL, Shirtcliff EA, Serbin LA, Fisher DB, Stack DM, Schwartzman AE (2011) Disentangling psychobiological mechanisms underlying internalizing and externalizing behaviors in youth: longitudinal and concurrent associations with cortisol. Horm Behav 59:123–132
Ryan BC, Vandenbergh JG (2002) Intrauterine position effects. Neurosci Biobehav Rev 26:665–678
Sagvolden T (2006) The alpha-2A adrenoceptor agonist guanfacine improves sustained attention and reduces overactivity and impulsiveness in an animal model of Attention-Deficit/Hyperactivity Disorder (ADHD). Behav Brain Funct 2:41
Sala M, Caverzasi E, Lazzaretti M, Morandotti N, De Vidovich G, Marraffini E, Gambini F, Isola M, De Bona M, Rambaldelli G, d'Allio G, Barale F, Zappoli F, Brambilla P (2011) Dorsolateral prefrontal cortex and hippocampus sustain impulsivity and aggressiveness in borderline personality disorder. J Affect Disord 131:417–421
Salas-Ramirez KY, Montalto PR, Sisk CL (2008) Anabolic androgenic steroids differentially affect social behaviors in adolescent and adult male Syrian hamsters. Horm Behav 53:378–385
Salas-Ramirez KY, Montalto PR, Sisk CL (2010) Anabolic steroids have long-lasting effects on male social behaviors. Behav Brain Res 208:328–335
Salminen M, Ravaja N (2008) Increased oscillatory theta activation evoked by violent digital game events. Neurosci Lett 435:69–72
Sandnabba NK, Lagerspetz KM, Jensen E (1994) Effects of testosterone exposure and fighting experience on the aggressive behavior of female and male mice selectively bred for intermale aggression. Horm Behav 28:219–231
Sano K, Mayanagi Y, Sekino H, Ogashiwa M, Ishijima B (1970) Results of stimulation and destruction of the posterior hypothalamus in man. J Neurosurg 33:689–707
Sarazin M, Michon A, Pillon B, Samson Y, Canuto A, Gold G, Bouras C, Dubois B, Giannakopoulos P (2003) Metabolic correlates of behavioral and affective disturbances in frontal lobe pathologies. J Neurol 250:827–833
Sasaki T, Uchida S (2012) The evolution of cooperation by social exclusion. Proc Biol Sci 280:20122498
Satpute AB, Wager TD, Cohen-Adad J, Bianciardi M, Choi JK, Buhle JT, Wald LL, Barrett LF (2013) Identification of discrete functional subregions of the human periaqueductal gray. Proc Natl Acad Sci U S A 110:17101–17106

Saxbe DE, Margolin G, Spies Shapiro LA, Baucom BR (2012) Does dampened physiological reactivity protect youth in aggressive family environments? Child Dev 83:821–830

Scerbo AS, Kolko DJ (1994) Salivary testosterone and cortisol in disruptive children: relationship to aggressive, hyperactive, and internalizing behaviors. J Am Acad Child Adolesc Psychiatry 33:1174–1184

Schaal B, Tremblay RE, Soussignan R, Susman EJ (1996) Male testosterone linked to high social dominance but low physical aggression in early adolescence. J Am Acad Child Adolesc Psychiatry 35:1322–1330

Schiffer B, Müller BW, Scherbaum N, Hodgins S, Forsting M, Wiltfang J, Gizewski ER, Leygraf N (2011) Disentangling structural brain alterations associated with violent behavior from those associated with substance use disorders. Arch Gen Psychiatry 68:1039–1049

Schjolden J, Basic D, Winberg S (2009) Aggression in rainbow trout is inhibited by both MR and GR antagonists. Physiol Behav 98:625–630

Schlinger BA, Callard GV (1989) Estrogen receptors in quail brain: a functional relationship to aromatase and aggresiveness. Biol Reprod 40:268–275

Schlinger BA, Callard GV (1990) Aromatization mediates aggressive behavior in quail. Gen Comp Endocrinol 79:39–53

Schmahl C, Bremner JD (2006) Neuroimaging in borderline personality disorder. J Psychiatr Res 40:419–427

Schneider F, Habel U, Kessler C, Posse S, Grodd W, Müller-Gärtner HW (2000) Functional imaging of conditioned aversive emotional responses in antisocial personality disorder. Neuropsychobiology 42:192–201

Schobert B, Lanyi JK (1982) Halorhodopsin is a light-driven chloride pump. J Biol Chem 257:10306–10313

Schubert MI, Porkess MV, Dashdorj N, Fone KC, Auer DP (2009) Effects of social isolation rearing on the limbic brain: a combined behavioral and magnetic resonance imaging volumetry study in rats. Neuroscience 159:21–30

Schuett GW, Grober MS (2000) Post-fight levels of plasma lactate and corticosterone in male copperheads, Agkistrodon contortrix (Serpentes, Viperidae): differences between winners and losers. Physiol Behav 71:335–341

Schuurman T (1980) Hormonal correlates of agonistic behaviour in adult male rats. Prog Brain Res 53:415–420

Schwartzer JJ, Ricci LA, Melloni RH Jr (2009) Interactions between the dopaminergic and GABAergic neural systems in the lateral anterior hypothalamus of aggressive AAS- treated hamsters. Behav Brain Res 203:15–22

Scordalakes EM, Rissman EF (2004) Aggression and arginine vasopressin immunoreactivity regulation by androgen receptor and estrogen receptor alpha. Genes Brain Behav 3:20–26

Scotti MA, Belén J, Jackson JE, Demas GE (2008) The role of androgens in the mediation of seasonal territorial aggression in male Siberian hamsters (Phodopus sungorus). Physiol Behav 95:633–640

Scotti MA, Schmidt KL, Newman AE, Bonu T, Soma KK, Demas GE (2009) Aggressive encounters differentially affect serum dehydroepiandrosterone and testosterone concentrations in male Siberian hamsters (Phodopus sungorus). Horm Behav 56:376–381

Seckl JR, Meaney MJ (2006) Glucocorticoid "programming" and PTSD risk. Ann NY Acad Sci 1071:351–378

Sekine Y, Ouchi Y, Takei N, Yoshikawa E, Nakamura K, Futatsubashi M, Okada H, Minabe Y, Suzuki K, Iwata Y, Tsuchiya KJ, Tsukada H, Iyo M, Mori N (2006) Brain serotonin transporter density and aggression in abstinent methamphetamine abusers. Arch Gen Psychiatry 63:90–100

Sellitto M, Ciaramelli E, di Pellegrino G (2011) The neurobiology of intertemporal choice: insight from imaging and lesion studies. Rev Neurosci 22:565–574

Serrien DJ, Ivry RB, Swinnen SP (2007) The missing link between action and cognition. Prog Neurobiol 82:95–107

Sesack SR, Deutch AY, Roth RH, Bunney BS (1989) Topographical organization of the efferent projections of the medial prefrontal cortex in the rat: an anterograde tract-tracing study with Phaseolus vulgaris leucoagglutinin. J Comp Neurol 290:213–242

Seward JP (1945) Aggressive behaviour in the rat: general characteristics, age and sex differences. J Comp Psychol 38:423–430

Sgoifo A, Pozzato C, Costoli T, Manghi M, Stilli D, Ferrari PF, Ceresini G, Musso E (2001) Cardiac autonomic responses to intermittent social conflict in rats. Physiol Behav 73:343–349

Shackman AJ, Salomons TV, Slagter HA, Fox AS, Winter JJ, Davidson RJ (2011) The integration of negative affect, pain and cognitive control in the cingulate cortex. Nat Rev Neurosci 12:154–167

Shaikh MB, Brutus M, Siegel HE, Siegel A (1984) Differential control of aggression by the midbrain. Exp Neurol 83:436–442

Shaikh MB, Barrett JA, Siegel A (1987) The pathways mediating affective defense and quiet biting attack behavior from the midbrain central gray of the cat: an autoradiographic study. Brain Res 437:9–25

Shaikh MB, Lu CL, Siegel A (1991) An enkephalinergic mechanism involved in amygdaloid suppression of affective defence behavior elicited from the midbrain periaqueductal gray in the cat. Brain Res 559:109–117

Shaikh MB, Steinberg A, Siegel A (1993) Evidence that substance P is utilized in medial amygdaloid facilitation of defensive rage behavior in the cat. Brain Res 625:283–294

Shamay-Tsoory SG, Harari H, Aharon-Peretz J (2010) Levkovitz Y The role of the orbitofrontal cortex in affective theory of mind deficits in criminal offenders with psychopathic tendencies. Cortex 46:668–677

Shehzad Z, DeYoung CG, Kang Y, Grigorenko EL, Gray JR (2012) Interaction of COMT val158met and externalizing behavior: relation to prefrontal brain activity and behavioral performance. Neuroimage 60:2158–2168

Sheng T, Gheytanchi A, Aziz-Zadeh L (2010) Default network deactivations are correlated with psychopathic personality traits. PLoS One 5:e12611

Shibata H, Naito J (2008) Organization of anterior cingulate and frontal cortical projections to the retrosplenial cortex in the rat. J Comp Neurol 506:30–45

Shigematsu N, Fukuda T, Yamamoto T, Nishioku T, Yamaguchi T, Himeno M, Nakayama KI, Tsukuba T, Kadowaki T, Okamoto K, Higuchi S, Yamamoto K (2008) Association of cathepsin E deficiency with the increased territorial aggressive response of mice. J Neurochem 105:1394–1404

Shimozuru M, Kikusui T, Takeuchi Y, Mori Y (2008) Effects of isolation-rearing on the development of social behaviors in male Mongolian gerbils (Meriones unguiculatus). Physiol Behav 94:491–500

Shirtcliff EA, Essex MJ (2008) Concurrent and longitudinal associations of basal and diurnal cortisol with mental health symptoms in early adolescence. Dev Psychobiol 50:690–703

Shirtcliff EA, Granger DA, Booth A, Johnson D (2005) Low salivary cortisol levels and externalizing behavior problems in youth. Dev Psychopathol 17:167–184

Shoal GD, Giancola PR, Kirillova GP (2003) Salivary cortisol, personality, and aggressive behavior in adolescent boys: a 5-year longitudinal study. J Am Acad Child Adolesc Psychiatry 42:1101–1107

Shufelt CL, Braunstein GD (2009) Safety of testosterone use in women. Maturitas 63:63–66

Siegel A, Douard J (2011) Who's flying the plane: serotonin levels, aggression and free will. Int J Law Psychiatry 34:20–29

Siegel A, Edinger HM (1983) Role of the limbic system in hypothalamically elicited attack behavior. Neurosci Biobehav Rev 7:395–407

Siegel A, Flynn JP (1968) Differential effects of electrical stimulation and lesions of the hippocampus and adjacent regions upon attack behavior in cats. Brain Res 7:252–267

Siegel A, Pott CB (1988) Neural substrates of aggression and flight in the cat. Prog Neurobiol 31:261–283

Siegel A, Edinger H, Dotto M (1975) Effects of electrical stimulation of the lateral aspect of the prefrontal cortex upon attack behavior in cats. Brain Res 93:473–484

Siegel A, Edinger H, Koo A (1977) Suppression of attack behavior in the cat by the prefrontal cortex: role of the mediodorsal thalamic nucleus. Brain Res 127:185–190

Siegel A, Brutus M (1990) Neural substrates of aggression and rage in the cat. In: Epstein AN, Morrison AR (eds) Progress in psychobiology and physiological psychology. Academic, San Diego, CA, pp 135–233

Siegel A, Schubert KL, Shaikh MB (1997) Neurotransmitters regulating defensive rage behavior in the cat. Neurosci Biobehav Rev 21:733–742

Siegel A, Roeling TA, Gregg TR, Kruk MR (1999) Neuropharmacology of brainstimulation-evoked aggression. Neurosci Biobehav Rev 23:359–389

Siegel A, Bhatt S, Bhatt R, Zalcman SS (2007) The neurobiological bases for development of pharmacological treatments of aggressive disorders. Curr Neuropharmacol 5:135–147

Siegrist J, Menrath I, Stöcker T, Klein M, Kellermann T, Shah NJ, Zilles K, Schneider F (2005) Differential brain activation according to chronic social reward frustration. Neuroreport 16:1899–1903

Siever LJ, Buchsbaum MS, New AS, Spiegel-Cohen J, Wei T, Hazlett EA, Sevin E, Nunn M, Mitropoulou V (1999) d,l-fenfluramine response in impulsive personality disorder assessed with [18F]fluorodeoxyglucose positron emission tomography. Neuropsychopharmacology 20:413–423

Sigurdsson T, Doyère V, Cain CK, LeDoux JE (2007) Long-term potentiation in the amygdala: a cellular mechanism of fear learning and memory. Neuropharmacology 52:215–227

Silverin B (1993) Territorial aggressiveness and its relation to the endocrine system in the pied flycatcher. Gen Comp Endocrinol 89:206–213

Simerly RB, Swanson LW, Gorski RA (1985) Reversal of the sexually dimorphic distribution of serotonin-immunoreactive fibers in the medial preoptic nucleus by treatment with perinatal androgen. Brain Res 340:91–98

Simmons RK, Howard JL, Simpson DN, Akil H, Clinton SM (2012) DNA methylation in the developing hippocampus and amygdala of anxiety-prone versus risk-taking rats. Dev Neurosci 34:58–67

Singh JP, Volavka J, Czobor P, Van Dorn RA (2012) A meta-analysis of the Val158Met COMT polymorphism and violent behavior in schizophrenia. PLoS One 7:e43423

Sinke CB, Sorger B, Goebel R, de Gelder B (1999) Tease or threat? Judging social interactions from bodily expressions. Neuroimage 49:1717–1727

Sinnamon HM (1990) Locomotor stepping elicited by electrical stimulation of the hypothalamus persists after lesion of descending fibers of passage. Physiol Behav 8:261–266

Sjoberg RL, Ducci F, Barr CS, Newman TK, Dell'osso L, Virkkunen M, Goldman D (2008) A non-additive interaction of a functional MAO-A VNTR and testosterone predicts antisocial behavior. Neuropsychopharmacology 33:425–430

Sluyter F, van Oortmerssen GA, de Ruiter AJ, Koolhaas JM (1996) Aggression in wild house mice: current state of affairs. Behav Genet 26:489–496

Sluyter F, Keijser JN, Boomsma DI, van Doornen LJ, van den Oord EJ, Snieder H (2000) Genetics of testosterone and the aggression-hostility-anger (AHA) syndrome: a study of middle-aged male twins. Twin Res 3:266–276

Sluyter F, Arseneault L, Moffitt TE, Veenema AH, de Boer SF, Koolhaas JM (2003) Toward an animalmodel for antisocial behavior: parallels between mice and humans. Behav Genet 33:563–574

Small IF, Heimburger RF, Small JG, Milstein V, Moore DF (1977) Follow-up of stereotaxic amygdalotomy for seizure and behavior disorders. Biol Psychiatry 12:401–411

Smith DA, Flynn JP (1980) Afferent projections to affective attack sites in cat hypothalamus. Brain Res 194:41–51

Smith DE, King MB, Hoebel BG (1970) Lateral hypothalamic control of killing: evidence for a cholinoceptive mechanism. Science 167:900–901

Snoek H, van Goozen SH, Matthys W, Sigling HO, Koppeschaar HP, Westenberg HG, van Engeland H (2002) Serotonergic functioning in children with oppositional defiant disorder: a sumatriptan challenge study. Biol Psychiatry 51:319–325

Soderstrom H, Blennow K, Forsman A, Liesivuori J, Pennanen S, Tiihonen J (2004) A controlled study of tryptophan and cortisol in violent offenders. J Neural Transm 111:1605–1610

Soler H, Vinayak P, Quadagno D (2000) Biosocial aspects of domestic violence. Psychoneuroendocrinology 25:721–739

Soloff PH, Meltzer CC, Becker C, Greer PJ, Kelly TM, Constantine D (2003) Impulsivity and prefrontal hypometabolism in borderline personality disorder. Psychiatry Res 123:153–163

Soloff PH, Meltzer CC, Becker C, Greer PJ, Constantine D (2005) Gender differences in a fenfluramine-activated FDG PET study of borderline personality disorder. Psychiatry Res 138:183–195

Soloff P, Nutche J, Goradia D, Diwadkar V (2008) Structural brain abnormalities in borderline personality disorder: a voxel-based morphometry study. Psychiatry Res 164:223–236

Solomon MB, Karom MC, Norvelle A, Markham CA, Erwin WD, Huhman KL (2009) Gonadal hormones modulate the display of conditioned defeat in male Syrian hamsters. Horm Behav 56:423–428

Soma KK (2006) Testosterone and aggression: berthold, birds and beyond. J Neuroendocrinol 18:543–551

Soma KK, Sullivan KA, Tramontin AD, Saldanha CJ, Schlinger BA, Wingfield JC (2000) Acute and chronic effects of an aromatase inhibitor on territorial aggression in breeding and nonbreeding male song sparrows. J Comp Physiol A 186:759–769

Soma KK, Schlinger BA, Wingfield JC, Saldanha CJ (2003) Brain aromatase, 5 alpha-reductase, and 5 beta-reductase change seasonally in wild male song sparrows: relationship to aggressive and sexual behavior. J Neurobiol 56:209–221

Soma KK, Scotti MA, Newman AE, Charlier TD, Demas GE (2008) Novel mechanisms for neuroendocrine regulation of aggression. Front Neuroendocrinol 29:476–489

Sondeijker FE, Ferdinand RF, Oldehinkel AJ, Tiemeier H, Ormel J, Verhulst FC (2008) HPA-axis activity as a predictor of future disruptive behaviors in young adolescents. Psychophysiology 45:398–404

Sonnen AEH, Manen JV, van Dijk B (1976) Results of amygdalotomy and fornicotomy in temporal lobe epilepsy and behavior disorders. Acta Neurochir Suppl 23:215–219

Spitzer M, Fischbacher U, Herrnberger B, Grön G, Fehr E (2007) The neural signature of social norm compliance. Neuron 56:185–196

Spoont MR, Kuskowski M, Pardo JV (2010) Autobiographical memories of anger in violent and non-violent individuals: a script-driven imagery study. Psychiatry Res 183:225–229

Stalenheim EG, Eriksson E, von Knorring L, Wide L (1998) Testosterone as a biological marker in psychopathy and alcoholism. Psychiatry Res 77:79–88

Stanworth RD, Jones TH (2008) Testosterone for the aging male; current evidence and recommended practice. Clin Interv Aging 3:25–44

Stefani MR, Groth K, Moghaddam B (2003) Glutamate receptors in the rat medial prefrontal cortex regulate set-shifting ability. Behav Neurosci 117:728–737

Steinert T (2002) Prediction of inpatient violence. Acta Psychiatr Scand Suppl 412:133–141

Sternson SM (2013) Hypothalamic survival circuits: blueprints for purposive behaviors. Neuron 77:810–824

Sterzer P, Stadler C, Krebs A, Kleinschmidt A, Poustka F (2005) Abnormal neural responses to emotional visual stimuli in adolescents with conduct disorder. Biol Psychiatry 57:7–15

Sterzer P, Stadler C, Poustka F, Kleinschmidt A (2007) A structural neural deficit in adolescents with conduct disorder and its association with lack of empathy. Neuroimage 37:335–342

Stewart J, Rajabi H (1994) Estradiol derived from testosterone in prenatal life affects the development of catecholamine systems in the frontal cortex in the male rat. Brain Res 646:157–160

Stivers JA, Skirboll LR, Long R, Crawley JN (1988) Anatomical analysis of frontal cortex sites at which carbachol induces motor seizures in the rat. Pharmacol Biochem Behav 30:129–136

Strenziok M, Krueger F, Heinecke A, Lenroot RK, Knutson KM, van der Meer E, Grafman J (2011) Developmental effects of aggressive behavior in male adolescents assessed with structural and functional brain imaging. Soc Cogn Affect Neurosci 6:2–11

Stupfel M, Pavely A (1990) Ultradian, circahoral and circadian structures in endothermic vertebrates and humans. Comp Biochem Physiol A Comp Physiol 96:1–11

Su T, Pagliaro M, Schmidt PJ, Pickar D, Wolkowitz O, Rubinow DR (1993) Neurospychiatric effects of anabolic steroids in male normal volunteers. J Am Med Assoc 269:2760–2764

Subramanian HH, Holstege G (2010) Periaqueductal gray control of breathing. Adv Exp Med Biol 669:353–358

Sugerman RA, Demski LS (1978) Agonistic behavior elicited by electrical stimulation of the brain in Western collared lizards, Crotaphytus collaris. Brain Behav Evol 15:446–469

Sugiura M, Wakusawa K, Sekiguchi A, Sassa Y, Jeong H, Horie K, Sato S, Kawashima R (2009) Extraction of situational meaning by integrating multiple meanings in a complex environment: a functional MRI study. Hum Brain Mapp 30:2676–2688

Summers CH, Summers TR, Moore MC, Korzan WJ, Woodley SK, Ronan PJ, Höglund E, Watt MJ, Greenberg N (2003) Temporal patterns of limbic monoamine and plasma corticosterone response during social stress. Neuroscience 116:553–563

Summers CH, Watt MJ, Ling TL, Forster GL, Carpenter RE, Korzan WJ, Lukkes JL, Overli O (2005) Glucocorticoid interaction with aggression in non-mammalian vertebrates: reciprocal action. Eur J Pharmacol 526:21–35

Sze PY, Iqbal Z (1994) Glucocorticoid action on depolarization-dependent calcium influx in brain synaptosomes. Neuroendocrinology 59:457–465

Takahashi LK, Gladstone CD (1988) Medial amygdaloid lesions and the regulation of sociosexual behavioral patterns across the estrous cycle in female golden hamsters. Behav Neurosci 102:268–275

Takahashi A, Shimamoto A, Boyson CO, DeBold JF, Miczek KA (2010) GABA(B) receptor modulation of serotonin neurons in the dorsal raphé nucleus and escalation of aggression in mice. J Neurosci 30:11771–11780

Takahashi A, Quadros IM, Miczek KA (2011) Brain serotonin receptors and transporters: initiation vs. termination of escalated aggression. Psychopharmacology 213:183–212

Takahashi A, Quadros IM, de Almeida RM, Miczek KA (2012) Behavioral and pharmacogenetics of aggressive behavior. Curr Top Behav Neurosci. PMID: 22297576 [Epub ahead of print]

Tanji J, Shima K (1996) Supplementary motor cortex in organization of movement. Eur Neurol 36 (Suppl 1):13–19

Taylor S (1967) Aggressive behavior and physiological arousal as a function of provocation and the tendency to inhibit aggression. J Pers 35:297–310

Tebartz van Elst L, Hesslinger B, Thiel T, Geiger E, Haegele K, Lemieux L, Lieb K, Bohus M, Hennig J, Ebert D (2003) Frontolimbic brain abnormalities in patients with borderline personality disorder: a volumetric magnetic resonance imaging study. Biol Psychiatry 54:163–171

Tekin S, Cummings JL (2002) Frontal-subcortical neuronal circuits and clinical neuropsychiatry: an update. J Psychosom Res 53:647–654

Terburg D, Morgan B, van Honk J (2009) The testosterone-cortisol ratio: a hormonal marker for proneness to social aggression. Int J Law Psychiatry 32:216–223

Terzian H, Ore GD (1955) Syndrome of Klüver and Bucy; reproduced in man by bilateral removal of the temporal lobes. Neurology 5:373–380, Cited based on Fountas and Smith, 2007

Tiihonen J, Rossi R, Laakso MP, Hodgins S, Testa C, Perez J, Repo-Tiihonen E, Vaurio O, Soininen H, Aronen HJ, Könönen M, Thompson PM, Frisoni GB (2008) Brain anatomy of persistent violent offenders: more rather than less. Psychiatry Res 163:201–212

Timmer M, Sandi C (2010) A role for glucocorticoids in the long-term establishment of a social hierarchy. Psychoneuroendocrinology 35:1543–1552

Tokarz RR (1987) Effects of corticosterone treatment on male aggressive behavior in a lizard (Anolis sagrei). Horm Behav 21:358–370

Tornatzky W, Miczek KA (1994) Behavioral and autonomic responses to intermittent social stress: differential protection by clonidine and metoprolol. Psychopharmacology (Berl) 116:346–356

Toth M, Halász J, Mikics E, Barsy B, Haller J (2008) Early social deprivation induces disturbed social communication and violent aggression in adulthood. Behav Neurosci 122:849–854

Toth M, Fuzesi T, Halasz J, Tulogdi A, Haller J (2010) Neural inputs of the hypothalamic "aggression area" in the rat. Behav Brain Res 215:7–20

Toth M, Mikics E, Tulogdi A, Aliczki M, Haller J (2011) Post-weaning social isolation induces abnormal forms of aggression in conjunction with increased glucocorticoid and autonomic stress responses. Horm Behav 60:28–36

Toth M, Tulogdi A, Biro L, Soros P, Mikics E, Haller J (2012) The neural background of hyper-emotional aggression induced by post-weaning social isolation. Behav Brain Res 233:120–129

Tourino C, Eban-Rothschild A, de Lecea L (2013) Optogenetics in psychiatric diseases. Curr Opin Neurobiol 23:430–435

Tracey I (2005) Nociceptive processing in the human brain. Curr Opin Neurobiol 15:478–487

Trainor BC, Marler CA (2001) Testosterone, paternal behavior, and aggression in the monogamous California mouse (Peromyscus californicus). Horm Behav 40:32–42

Trainor BC, Bird IM, Marler CA (2004) Opposing hormonal mechanisms of aggression revealed through short-lived testosterone manipulations and multiple winning experiences. Horm Behav 45:115–121

Trainor BC, Greiwe KM, Nelson RJ (2006) Individual differences in estrogen receptor alpha in select brain nuclei are associated with individual differences in aggression. Horm Behav 50:338–345

Trainor BC, Finy MS, Nelson RJ (2008a) Paternal aggression in a biparental mouse: parallels with maternal aggression. Horm Behav 53:200–207

Trainor BC, Finy MS, Nelson RJ (2008b) Rapid effects of estradiol on male aggression depend on photoperiod in reproductively non-responsive mice. Horm Behav 53:192–199

Tranel D, Bechara A, Denburg NL (2002) Asymmetric functional roles of right and left ventromedial prefrontal cortices in social conduct, decision-making, and emotional processing. Cortex 38:589–612

Tremblay RE (1999) When children's social development fails. In: Keating DP, Hertzman C (eds) Developmental health and the wealth of nations: social, biological, and educational dynamics. The Guilford Press, New York, NY, pp 55–71

Tremblay RE (2008) Understanding development and prevention of chronic physical aggression: towards experimental epigenetic studies. Philos Trans R Soc Lond B Biol Sci 363:2613–2622

Trentacosta CJ, Shaw DS (2008) Maternal predictors of rejecting parenting and early adolescent antisocial behavior. J Abnorm Child Psychol 36:247–259

Trifonova ST, Gantenbein M, Turner JD, Muller CP (2013) The use of saliva for assessment of cortisol pulsatile secretion by deconvolution analysis. Psychoneuroendocrinology 38(7):1090–1101

Tsoory MM, Vouimba RM, Akirav I, Kavushansky A, Avital A, Richter-Levin G (2008) Amygdala modulation of memory-related processes in the hippocampus: potential relevance to PTSD. Prog Brain Res 167:35–51

Tsuda MC, Yamaguchi N, Ogawa S (2011) Early life stress disrupts peripubertal development of aggression in male mice. Neuroreport 22:259–263

Tulogdi A, Toth M, Halasz J, Mikics E, Fuzesi T, Haller J (2010) Brain mechanisms involved in predatory aggression are activated in a laboratory model of violent intra-specific aggression. Eur J Neurosci 32:1744–1753

Tulogdi A, Toth M, Barsvári B, Biró L, Mikics E, Haller J (2014) Effects of resocialization on post-weaning social isolation-induced abnormal aggression and social deficits in rats. Dev Psychobiol 56(1):49–57

Turner AK (1994) Genetic and hormonal influences on male violence. In: Archer J (ed) Male violence. Routledge, New York, NY, pp 233–252

Ulrich-Lai YM, Herman JP (2009) Neural regulation of endocrine and autonomic stress responses. Nat Rev Neurosci 10:397–409

Vaillancourt T, Sunderani S (2011) Psychopathy and indirect aggression: the roles of cortisol, sex, and type of psychopathy. Brain Cogn 77:170–175
Vaillancourt KL, Dinsdale NL, Hurd PL (2012) Estrogen receptor 1 promoter polymorphism and digit ratio in men. Am J Hum Biol 24:682–689
Valenstein ES, Cox VC, Kakolewski JW (1968) Modification of motivated behavior elicited by electrical stimulation of the hypothalamus. Science 159(3819):1119–1121
van Anders SM, Goldey KL, Kuo PX (2011) The steroid/peptide theory of social bonds: integrating testosterone and peptide responses for classifying social behavioral contexts. Psychoneuroendocrinology 36:1265–1275
van Bokhoven I, Van Goozen SH, van Engeland H, Schaal B, Arseneault L, Séguin JR, Nagin DS, Vitaro F, Tremblay RE (2005) Salivary cortisol and aggression in a population-based longitudinal study of adolescent males. J Neural Transm 112:1083–1096
Van de Wiel N, van Goozen S, Matthys W, Snoek H, van Engeland H (2004) Cortisol and treatment effect in children with disruptive behavior disorders: a preliminary study. J Am Acad Child Adoloscent Psychiatry 43:1011–1018
Van den Bergh BR, Van Calster B, Pinna Puissant S, Van Huffel S (2008) Self-reported symptoms of depressed mood, trait anxiety and aggressive behavior in post-pubertal adolescents: associations with diurnal cortisol profiles. Horm Behav 54:253–257
van der Meij L, Almela M, Buunk AP, Dubbs S, Salvador A (2012) 2D:4D in men is related to aggressive dominance but not to sociable dominance. Aggress Behav 38:208–212
van Goozen SH, Matthys W, Cohen-Kettenis PT, Gispen-de Wied C, Wiegant VM, van Engeland H (1998a) Salivary cortisol and cardiovascular activity during stress in oppositional-defiant disorder boys and normal controls. Biol Psychiatry 43:531–539
van Goozen SH, Matthys W, Cohen-Kettenis PT, Thijssen JH, van Engeland H (1998b) Adrenal androgens and aggression in conduct disorder prepubertal boys and normal controls. Biol Psychiatry 43:156–158
van Goozen SH, Matthys W, Cohen-Kettenis PT, Buitelaar JK, van Engeland H (2000a) Hypothalamic-pituitary-adrenal axis and autonomic nervous system activity in disruptive children and matched controls. J Am Acad Child Adolesc Psychiatry 39:1438–1445
van Goozen SHM, van den Ban E, Matthys W, Cohen-Kettenis PT, Thijssen JHH, van Engeland H (2000b) Increased adrenal androgen functioning in children with oppositional defiant disorder: a comparison with psychiatric and normal controls. J Am Acad Child Adolesc Psychiatry 39:1446–1451
van Goozen SH, Fairchild G, Snoek H, Harold GT (2007) The evidence for a neurobiological model of childhood antisocial behavior. Psychol Bull 133:149–182
van Honk J, Harmon-Jones E, Morgan BE, Schutter DJ (2010) Socially explosive minds: the triple imbalance hypothesis of reactive aggression. J Pers 78:67–94
van Manen J, van Veelen CW (1988) Experiences in psycho-surgery in the Netherlands. Acta Neurochir Suppl 44:167–169
van Oortmerssen GA, Dijk DJ, Schuurman T (1987) Studies in wild house mice. II. Testosterone and aggression. Horm Behav 21:139–152
van Oortmerssen GA, Benus RF, Sluyter F (1992) Studies on wild house mice. IV. On the heredity of testosterone and readiness to attack. Aggress Behav 18:143–148
van Staaden MJ, Searcy WA, Hanlon RT (2011) Signaling aggression. Adv Genet 75:23–49
Vanyukov MM, Moss HB, Plail JA, Blackson T, Mezzich AC, Tarter RE (1993) Antisocial symptoms in preadolescent boys and in their parents: associations with cortisol. Psychiatry Res 46:9–17
Veenema AH (2009) Early life stress, the development of aggression and neuroendocrine and neurobiological correlates: what can we learn from animal models? Front Neuroendocrinol 30:497–518
Veenema AH (2012) Toward understanding how early-life social experiences alter oxytocin- and vasopressin-regulated social behaviors. Horm Behav 61:304–312
Veenema AH, Neumann ID (2008) Central vasopressin and oxytocin release: regulation of complex social behaviours. Prog Brain Res 170:261–276

Veenema AH, Neumann ID (2009) Maternal separation enhances offensive play-fighting, basal corticosterone and hypothalamic vasopressin mRNA expression in juvenile male rats. Psychoneuroendocrinology 34:463–467

Veenema AH, Meijer OC, de Kloet ER, Koolhaas JM, Bohus BG (2003) Differences in basal and stress-induced HPA regulation of wild house mice selected for high and low aggression. Horm Behav 43:197–204

Veenema AH, Koolhaas JM, de Kloet ER (2004) Basal and stress-induced differences in HPA axis, 5-HT responsiveness, and hippocampal cell proliferation in two mouse lines. Ann NY Acad Sci 1018:255–265

Veenema AH, Blume A, Niederle D, Buwalda B, Neumann ID (2006) Effects of early life stress on adult male aggression and hypothalamic vasopressin and serotonin. Eur J Neurosci 24:1711–1720

Veenema AH, Bredewold R, Neumann ID (2007a) Opposite effects of maternal separation on intermale and maternal aggression in C57BL/6 mice: link to hypothalamic vasopressin and oxytocin immunoreactivity. Psychoneuroendocrinology 32:437–450

Veenema AH, Torner L, Blume A, Beiderbeck DI, Neumann ID (2007b) Low inborn anxiety correlates with high intermale aggression: link to ACTH response and neuronal activation of the hypothalamic paraventricular nucleus. Horm Behav 51:11–19

Veenema AH, Beiderbeck DI, Lukas M, Neumann ID (2010) Distinct correlations of vasopressin release within the lateral septum and the bed nucleus of the stria terminalis with the display of intermale aggression. Horm Behav 58:273–281

Veening JG, Coolen LM, de Jong TR, Joosten HW, de Boer SF, Koolhaas JM, Olivier B (2005) Do similar neural systems subserve aggressive and sexual behaviour in male rats? Insights from c-Fos and pharmacological studies. Eur J Pharmacol 526:226–239

Veenstra-VanderWeele J, Anderson GM, Cook EH Jr (2000) Pharmacogenetics and the serotonin system: initial studies and future directions. Eur J Pharmacol 410:165–181

Veit R, Lotze M, Sewing S, Missenhardt H, Gaber T, Birbaumer N (2010) Aberrant social and cerebral responding in a competitive reaction time paradigm in criminal psychopaths. Neuroimage 49:3365–3372

Velley L, Verney C, Kempf E, Berger B (1988) Opposite effects of ibotenic acid and 6-hydroxydopamine lesions of the lateral hypothalamus on intracranial self-stimulation and stimulation-induced locomotion. Neuroscience 25:595–603

Vergnes M, Karli P (1969) Interspecific rat-mice agression behavior: effects of electric stimulation of the lateral hypothalamus, amygdala and hypopocampus. J Physiol 61(Suppl 2):425

Verona E, Kilmer A (2007) Stress exposure and affective modulation of aggressive behavior in men and women. J Abnormal Psychol 116:410–421

Verwer RW, Van Vulpen EH, Van Uum JF (1996) Postnatal development of amygdaloid projections to the prefrontal cortex in the rat studied with retrograde and anterograde tracers. J Comp Neurol 376:75–96

Victoroff J, Quota S, Adelman JR, Celinska B, Stern N, Wilcox R, Sapolsky RM (2011) Support for religio-political aggression among teenaged boys in Gaza: Part II: Neuroendocrinological findings. Aggress Behav 37:121–132

Virkkunen M (1985) Urinary free cortisol secretion in habitually violent offenders. Acta Psychiatr Scand 72:40–44

Virkkunen M, Kallio E, Rawlings R, Tokola R, Poland RE, Guidotti A, Nemeroff C, Bissette G, Kalogeras K, Karonen SL, Linnoila M (1994a) Personality profiles and state aggressiveness in Finnish alcoholic violent offenders fire setters and healthy volunteers. Arch Gen Psychiatry 51:28–33

Virkkunen M, Rawlings R, Tokola R, Poland RE, Guidotti A, Nemeroff C, Bissette G, Kalogeras K, Karonen SL, Linnoila M (1994b) CSF biochemistries, glucose metabolism, and diurnal activity rhythms in alcoholic, violent offenders, fire setters, and healthy volunteers. Arch Gen Psychiatry 51:20–27

Vizi ES (1984) Non-synaptic interactions between neurons: modulation of neurochemical transmission. Pharmacological and clinical aspects. Wiley, New York, NY

Vizi ES (2000) Role of high-affinity receptors and membrane transporters in nonsynaptic communication and drug action in the central nervous system. Pharmacol Rev 52:63–89

Vizi ES, Fekete A, Karoly R, Mike A (2010) Non-synaptic receptors and transporters involved in brain functions and targets of drug treatment. Br J Pharmacol 160:785–809

Vochteloo JD, Koolhaas JM (1987) Medial amygdala lesions in male rats reduce aggressive behavior: interference with experience. Physiol Behav 41:99–102

Volkow ND, Tancredi LR, Grant C, Gillespie H, Valentine A, Mullani N, Wang GJ, Hollister L (1995) Brain glucose metabolism in violent psychiatric patients: a preliminary study. Psychiatry Res 61:243–253

Völlm B, Richardson P, McKie S, Reniers R, Elliott R, Anderson IM, Williams S, Dolan M, Deakin B (2010) Neuronal correlates and serotonergic modulation of behavioural inhibition and reward in healthy and antisocial individuals. J Psychiatr Res 44:123–131

Voracek M (2013) Differential correlations of digit ratio (2D:4D) with aggressive dominance and sociable dominance are not demonstrated: commentary on van der Meij, Almela, Buunk, Dubbs, and Salvador (2012, Aggressive Behavior, 38(3), 208–212). Aggress Behav 39:85–87

Voracek M, Stieger S (2009) Replicated nil associations of digit ratio (2D:4D) and absolute finger lengths with implicit and explicit measures of aggression. Psicothema 21:382–389

Walker WH (2003) Nongenomic actions of androgen in Sertoli cells. Curr Top Dev Biol 56:25–53

Wall VL, Fischer EK, Bland ST (2012) Isolation rearing attenuates social interaction-induced expression of immediate early gene protein products in the medial prefrontal cortex of male and female rats. Physiol Behav 107:440–450

Wallinga AE, ten Voorde AM, de Boer SF, Koolhaas JM, Buwalda B (2009) MDMA-induced serotonergic neurotoxicity enhances aggressiveness in low- but not high-aggressive rats. Eur J Pharmacol 618:22–27

Wang Z, Hulihan TJ, Insel TR (1997) Sexual and social experience is associated with different patterns of behavior and neural activation in male prairie voles. Brain Res 767:321–332

Wang D, Szyf M, Benkelfat C, Provençal N, Turecki G, Caramaschi D, Côté SM, Vitaro F, Tremblay RE, Booij L (2012a) Peripheral SLC6A4 DNA methylation is associated with in vivo measures of human brain serotonin synthesis and childhood physical aggression. PLoS One 7:e39501

Wang D, Zhang J, Zhang Z (2012b) Effect of testosterone and melatonin on social dominance and agonistic behavior in male Tscheskia triton. Behav Processes 89:271–277

Wang Y, He Z, Zhao C, Li L (2013) Medial amygdala lesions modify aggressive behavior and immediate early gene expression in oxytocin and vasopressin neurons during intermale exposure. Behav Brain Res 245:42–49

Warnault V, Darcq E, Levine A, Barak S, Ron D (2013) Chromatin remodeling – a novel strategy to control excessive alcohol drinking. Transl Psychiatry 3:e231

Warne GL, Faiman C, Reyes FI, Winter JSD (1977) Studies on human sexual development. V. Concentrations of testosterone, 17-hydroxyprogesterone and progesterone in human amniotic fluid throughout gestation. J Clin Endocrinol Metab 44:934–938

Watson TC, Jones MW, Apps R (2009) Electrophysiological mapping of novel prefrontal – cerebellar pathways. Front Integr Neurosci 3:18

Weerts EM, Miller LG, Hood KE, Miczek KA (1992) Increased GABAA-dependent chloride uptake in mice selectively bred for low aggressive behavior. Psychopharmacology (Berl) 108:196–204

Weiger WA, Bear DM (1988) An approach to the neurology of aggression. J Psychiatr Res 22:85–98

Weinberg A, Klonsky ED, Hajcak G (2009) Autonomic impairment in borderline personality disorder: a laboratory investigation. Brain Cogn 71:279–286

Weiner CL, Primeau M, Ehrmann DA (2004) Androgens and mood dysfunction in women: comparison of women with polycystic ovarian syndrome to healthy controls. Psychosom Med 66:356–362

Weiss CS, Coughlin JP (1979) Maintained aggressive behavior in gonadectomized male Siamese fighting fish (Betta splendens). Physiol Behav 23:173–177

Weissenberger AA, Dell ML, Liow K, Theodore W, Frattali CM, Hernandez D, Zametkin AJ (2001) Aggression and psychiatric comorbidity in children with hypothalamic hamartomas and their unaffected siblings. J Am Acad Child Adolesc Psychiatry 40:696–703

Wen JC, Hotchkiss AK, Demas GE, Nelson RJ (2004) Photoperiod affects neuronal nitric oxide synthase and aggressive behaviour in male Siberian hamsters (Phodopus sungorus). J Neuroendocrinol 16:916–921

White TL, Grover VK, de Wit H (2006) Cortisol effects of D-amphetamine relate to traits of fearlessness and aggression but not anxiety in healthy humans. Pharmacol Biochem Behav 85:123–131

White SF, Brislin SJ, Sinclair S, Blair JR (2013) Punishing unfairness: rewarding or the organization of a reactively aggressive response? Hum Brain Mapp. doi:10.1002/hbm.22316 [Epub ahead of print]

Whittle S, Allen NB, Lubman DI, Yücel M (2006) The neurobiological basis of temperament: towards a better understanding of psychopathology. Neurosci Biobehav Rev 30:511–525

Widom CS (1989) Child abuse, neglect, and adult behavior: research design and findings on criminality, violence, and child abuse. Am J Orthopsychiatry 59:355–367

Williams JR, Ghandour RM, Kub JE (2008) Female perpetration of violence in heterosexual intimate relationships: adolescence through adulthood. Trauma Violence Abuse 9:227–249

Windle RC, Windle M (1995) Longitudinal patterns of physical aggression: associations with adult social, psychiatric, and personality functioning and testosterone levels. Dev Psychopathol 7:563–585

Wingfield JC, Hegner RE, Dufty AM Jr, Ball GF (1990) The "challenge hypothesis": theoretical implications for patterns of testosterone secretion, mating systems, and breeding strategies. Am Naturalist 136:829–846

Winslow JT, Miczek KA (1988) Androgen dependency of alcohol effects on aggressive behavior: a seasonal rhythm in high-ranking squirrel monkeys. Psychopharmacology (Berl) 95:92–98

Winslow JT, Ellingboe J, Miczek KA (1988) Effects of alcohol on aggressive behavior in squirrel monkeys: influence of testosterone and social context. Psychopharmacology (Berl) 95:356–363

Winstanley CA, Olausson P, Taylor JR, Jentsch JD (2010) Insight into the relationship between impulsivity and substance abuse from studies using animal models. Alcohol Clin Exp Res 34:1306–1318

Wirth MM, Schultheiss OC (2007) Basal testosterone moderates responses to anger faces in humans. Physiol Behav 90:496–505

Wise SP, Boussaoud D, Johnson PB, Caminiti R (1997) Premotor and parietal cortex: corticocortical connectivity and combinatorial computations. Annu Rev Neurosci 20:25–42

Witte AV, Flöel A, Stein P, Savli M, Mien LK, Wadsak W, Spindelegger C, Moser U, Fink M, Hahn A, Mitterhauser M, Kletter K, Kasper S, Lanzenberger R (2009) Aggression is related to frontal serotonin-1A receptor distribution as revealed by PET in healthy subjects. Hum Brain Mapp 30:2558–2570

Wommack JC, Salinas A, Melloni RH Jr, Delville Y (2004) Behavioural and neuroendocrine adaptations to repeated stress during puberty in male golden hamsters. J Neuroendocrinol 16:767–775

Wood GE, Young LT, Reagan LP, McEwen BS (2003) Acute and chronic restraint stress alter the incidence of social conflict in male rats. Horm Behav 43:205–213

Woodley SK, Matt KS, Moore MC (2000) Neuroendocrine responses in free-living female and male lizards after aggressive interactions. Physiol Behav 71:373–381

Woodman D, Hinton J (1978) Catecholamine balance during stress anticipation: an abnormality in maximum security hospital patients. J Psychosom Res 22:477–483

Woodworth CH (1971) Attack elicited in rats by electrical stimulation of the lateral hypothalamus. Physiol Behav 6:345–353

Woolley DG, Laeremans A, Gantois I, Mantini D, Vermaercke B, Op de Beeck HP, Swinnen SP, Wenderoth N, Arckens L, D'Hooge R (2013) Homologous involvement of striatum and prefrontal cortex in rodent and human water maze learning. Proc Natl Acad Sci U S A 110:3131–3136

Wu MV, Manoli DS, Fraser EJ, Coats JK, Tollkuhn J, Honda S, Harada N, Shah NM (2009) Estrogen masculinizes neural pathways and sex-specific behaviors. Cell 139:61–72

Yadav R, Gupta SC, Hillman BG, Bhatt JM, Stairs DJ, Dravid SM (2012) Deletion of glutamate delta-1 receptor in mouse leads to aberrant emotional and social behaviors. PLoS One 7: e32969

Yang Y, Raine A (2009) Prefrontal structural and functional brain imaging findings in antisocial, violent, and psychopathic individuals: a meta-analysis. Psychiatry Res 174:81–88

Yang EJ, Wilczynski W (2002) Relationships between hormones and aggressive behavior in green anole lizards: an analysis using structural equation modeling. Horm Behav 42:192–205

Yang Y, Raine A, Lencz T, Bihrle S, Lacasse L, Colletti P (2005) Prefrontal white matter in pathological liars. Br J Psychiatry 187:320–325

Yang SJ, Shin DW, Noh KS, Stein MA (2007) Cortisol is inversely correlated with aggression for those boys with attention deficit hyperactivity disorder who retain their reactivity to stress. Psychiatry Res 153:55–60

Yang Y, Glenn AL, Raine A (2008) Brain abnormalities in antisocial individuals: implications for the law. Behav Sci Law 26:65–83

Yang Y, Raine A, Narr KL, Colletti P, Toga AW (2009) Localization of deformations within the amygdala in individuals with psychopathy. Arch Gen Psychiatry 66:986–994

Yao R, Rameshwar P, Gregg T, Siegel A (2001) Co-localization of NK(1)-receptor mRNA with glutamate immunopositivity in cat hypothalamic neurons by the combination of in situ hybridization and immunohistochemistry. Brain Res Brain Res Protoc 7:154–161

Yildirim BO, Derksen JJ (2012a) A review on the relationship between testosterone and the interpersonal/affective facet of psychopathy. Psychiatry Res 197:181–198

Yildirim BO, Derksen JJ (2012b) A review on the relationship between testosterone and life-course persistent antisocial behavior. Psychiatry Res 200:984–1010

Yodyingyuad U, de la Riva C, Abbott DH, Herbert J, Keverne EB (1985) Relationship between dominance hierarchy, cerebrospinal fluid levels of amine transmitter metabolites (5-hydroxyindole acetic acid and homovanillic acid) and plasma cortisol in monkeys. Neuroscience 16:851–858

Yohe LR, Suzuki H, Lucas LR (2012) Aggression is suppressed by acute stress but induced by chronic stress: immobilization effects on aggression, hormones, and cortical 5-HT(1B)/striatal dopamine D(2) receptor density. Cogn Affect Behav Neurosci 12:446–459

Yu YZ, Shi JX (2009) Relationship between levels of testosterone and cortisol in saliva and aggressive behaviors of adolescents. Biomed Environ Sci 22:44–49

Zagrodzka J, Fonberg E (1977) Amygdalar area involved in predatory behavior in cats. Acta Neurobiol Exp 37:131–136

Zagrodzka J, Fonberg E (1978) Predatory versus alimentary behavior after amygdala lesions in cats. Physiol Behav 20:523–531

Zai CC, Muir KE, Nowrouzi B, Shaikh SA, Choi E, Berall L, Trépanier MO, Beitchman JH, Kennedy JL (2012) Possible genetic association between vasopressin receptor 1B and child aggression. Psychiatry Res 200:784–788

Zalcman SS, Siegel A (2006) The neurobiology of aggression and rage: role of cytokines. Brain Behav Immun 20:507–514

Zillman D (1979) Hostility and aggression. Erlbaum, Hillsdale, NJ

Zimmermann N, Zschocke J, Perisic T, Yu S, Holsboer F, Rein T (2012) Antidepressants inhibit DNA methyltransferase 1 through reducing G9a levels. Biochem J 448:93–102

Index

J. Haller, *Neurobiological Bases of Abnormal Aggression and Violent Behaviour*, DOI 10.1007/978-3-7091-1268-7, © Springer-Verlag Wien 2014

MIX
Papier aus verantwortungsvollen Quellen
Paper from responsible sources
FSC® C105338

If you have any concerns about our products,
you can contact us on
ProductSafety@springernature.com

In case Publisher is established outside the EU,
the EU authorized representative is:
Springer Nature Customer Service Center GmbH
Europaplatz 3, 69115 Heidelberg, Germany

Printed by Libri Plureos GmbH
in Hamburg, Germany